European Communities Oil and Gas Technological Development Projects

Second Status Report

European Communities Oil and Gas Technological Development Projects

Second Status Report

Compiled by
E Millich
J P Joulia
D Van Asselt

Commission of the European Communities
Directorate-General for Energy,
Brussels

Published by
Graham & Trotman
for the Commission of the European Communities

Published in 1984 by
Graham & Trotman Limited
Sterling House, 66 Wilton Road
London SW1V 1DE, United Kingdom

for the Commission of the European Communities, Directorate-General Information Market and Innovation

EUR 9549

ISBN 0 86010 616 0

Printed and bound in Great Britain

CONTENTS

P R E F A C E

The security of the Community's hydrocarbon supplies depends heavily on the efforts made by the European petroleum industry in oil and natural gas exploration and production.

To assist their efforts, the Community adopted, in 1973, a financial instrument (Regulation 3056/73) designed to encourage the development of new technologies for the exploration, exploitation, transport and storage of hydrocarbons.

These activities are carried out under particular difficult conditions in Europe. Improving technologies to make them more economic and better adapted to these conditions is, therefore, of special importance.

The present report sets out the results achieved so far by the Community support programme. Its publication coincides with the second Symposium on New Oil Technologies, where the most important results of the Community programme will be presented.

These results testify to the dynamism of European enterprises and their capacity to meet the technological challenge associated with the security of hydrocarbon supplies in the Community.

E. Davignon

INTRODUCTION

The Commission of the European Community has, by means of the Directorate General for Energy, been involved in energy research aimed at improving the energy supply situation of the Community. This involvement is on two levels, firstly, the Community supports research and development aimed at improving the technologies associated with the location and production of traditional fuels and, secondly, the Community is actively involved in research to replace traditional energy sources with suitable alternatives.

Given the parlous state of the energy supply situation in the Community, it was felt that a special effort was required to develop new technologies associated with improving the supply of traditional fuels and in developing and establishing alternative sources of energy.

The initiative of the Community was begun in 1973 when the Council approved Regulation (EEC) 3056/73 setting up a series of three-year research and development programmes in the oil and gas sector. This programme was one factor in the Community's response to the supply crisis of 1973.

The purpose of this report is to present the research and development carried out under contract in the framework of this programme, which is aimed at the improvement of the oil and gas supply situation by developing these technologies associated with the location, production and transportation of oil and gas. The report begins by examining the oil and gas programme in the context of energy research and development within the Directorate General for Energy. It then presents a summary of the programme's content, its implementation and supervision structures and the results to date. Finally, a detailed summary of each of the contracts concluded up to the eighth round of projects (1982). This summary, which constitutes the main part of this report, is presented by subject area.

COMMUNITY ENERGY RESEARCH AND DEVELOPMENT STRATEGY

The energy strategy of the Community has been developed in the context of the fact that 60% of the energy supplies of the Community are imported. The strategy has therefore evolved on three levels: firstly by endeavouring to secure traditional energy supplies for the Community, secondly by embarking on an energy saving programme aimed at reducing the demand for energy within the Community without at the same time causing economic disruption and thirdly by developing alternative energy sources which may in time supplant traditional forms of energy.

These considerations also influence the development of the Community's energy research and development programmes, which, in the case of the Directorate General for Energy, can be divided into four strategic sectors:

1) The improvement of supply of conventional energy sources by research and development in hydrocarbon technology.

2) The promotion of technological advance in the exploitation of alternative energy sources.

3) The identification and demonstration of technologies in the field of energy saving and oil substitution.

4) The improvement, by an active exploration programme, of the Community's supply of uranium.

The hydrocarbon technological development programme has been concentrated on developing techniques which should improve the supply situation of the Community in the shortest possible time. A good example of this type of technology is enhanced recovery, the development of which is already having a considerable impact on the ultimate recoverable reserved of crude oil, both within the Community and world wide.

THE COUNCIL PROGRAMME DECISION

The decision of the Council adopting Regulation (EEC) 3056/73 was taken on 9 November 1973. This Regulation set up a series of three-year research and development projects in the hydrocarbon sector.

Article 1 of the Regulation aptly describes the purpose of the programme:

"The Community may, in accordance with the conditions laid down hereinafter, grant financial support, in so far as this is essential, for the carrying out of projects (Community projects) which are of fundamental importance in ensuring the Community's suply of hydrocarbons."

The projects carried out thus far have been subdivided into the following fields:

01 - Geophysics and prospecting

02 - Drilling

03 - Production systems

05 - Secondary and enhanced recovery

06 - Environmental influence on offshore structures

07 - Auxiliary ship, submersibles and navigation systems

09 - Pipelaying

10 - Transportation

12 - Natural gas technology

13 - Energy sources

14 - Storage

15 - Miscellaneous

The amount of support allocated is generally expressed in accordance with the chances of success of the projects and their relative importance for the Community. Projects that, by exploitation of the results, will mean an increase in resource and/or an acceleration in the exploration and exploitation of hydrocarbon resources in the Community, have the benefit of the maximum rate of support of 40 percent. Transport and storage projects usually receive support of 30 to 35 percent.

Table 1 shows the sectorial distribution and the global funding situation of the first eight rounds of projects.

IMPLEMENTATION AND SUPERVISION STRUCTURES

The Commission has several responsibilities in the implementation of Regulation (EEC) 3056/73. Firstly, the Commission is responsible for issuing the call for tenders and for examining the proposals submitted. A proposal is then drawn up and submitted for discussion within the Energy Working Group of the Council. When a Council decision is taken, the services of the Commission are then responsible for the negotiation and conclusion of the contracts and the monitoring of the projects, both technically and financially during its lifetime. The technical coordination of the programme is the responsibility of Directorate C "Hydrocarbons", while the Contracts Division of Directorate A is responsible for all administrative matters.

After the signature of the contract, the projects are monitored both by examination of the periodic reports and by making annual technical and financial controls on site.

Annual reports concerning the state of advancement of the programme are made to the European Parliament and the Council.

Since the requirements for research and development in the oil and gas sector are constantly changing, the criteria for examining proposals are constantly reviewed. When a major change in criteria is proposed, a review meeting is held with the various national experts associated with the Energy Working Group of the Council.

STATUS OF IMPLEMENTATION AND FIRST RESULTS

A subsidy has been granted to 319 projects. Many of the projects have been completed but, as can be seen from the summaries, many are still in progress.

In terms of technical results, the programme has already contributed to the solution of many technical problems which were considered by experts in the field as bottlenecks.

While the summaries contain details of the results of individual projects, the following sections give some indication of the global situation.

- 1975 - 1st Round Projects

  The most significant results were obtained in the areas of pipelaying, in deep sea drilling and in deep water production techniques.

The pipelaying tests in the Straits of Messina and the Sicily Channel carried out by SNAM were directly responsible for proving the viability of deep water pipelaying in the Mediterranean and have led to the construction of the gas line from Algeria to Italy which will carry 12 billion m3 of natural gas per year.

In deep sea drilling, several new pieces of equipment, especially the drilling equipment of the drillship "Petrel", were developed by GERTH (Groupement européen de Recherches technologiques sur les Hydrocarbures). This development included major advances in deep sea blow-out preventer controls, re-entry sonar and various drilling tools. This project has led directly to the ability of Community companies to drill in up to 1,500 m water depths.

In deep sea production, studies were carried out by GERTH, which led to the construction of the Grondin underwater test station. This was the first test made on a live field of a totally underwater production system designed for deep water.

- 1976 - 2nd Round Projects

In deep sea drilling, support was given to develop several novel aspects of the dynamically positioned drillship "Pèlerin" (GERTH) and "Ben Ocean Lancer" (Ben Odeco). The development of these two vessels added considerably to the ability of Community companies to drill for hydrocarbons in a deep water hostile environment.

The second round of projects marked the beginning of interest in novel production systems. As a result of work done in this round, one Community contractor, Vickers Offshore, has been involved in the design of the world's first tethered buoyant platform for the Hutton Field in the North Sea. Also, another Community company, British Petroleum, is developing a deep water gas production system.

Excellent results were obtained in the area of natural gas technology and Community support has succeeded in putting Preussag and Salzgitter in the forefront of LNG developments offshore.

This round also marked the emergence of enhanced recovery technology, and a very successful pilot project was carried out by GERTH on the Châteaurenard Reservoir. If the results of this project prove applicable on a wide scale, the recoverable reserves of the Community will be substantially increased; estimates of the increase range from 10 to 50 million m3 of oil.

- 1977 - 3rd Round Projects

In production technology, a major project carried out by GERTH has led to the development and testing of components for a deep water production system. Every aspect of the system has been brought to a stage where a suitable field is now being sought on which to confirm the results of component tests. A valuable offshoot of this programme has been the further development of the J-curve pipelaying method, which is of major importance.

Preussag continued to develop their experience in gas technology by carrying out a comprehensive examination of offshore liquefaction of natural gas; a novel aspect of this programme is the development of an offshore LNG loading arm. This project will have many applications in North Sea marginal field development.

Three enhanced recovery programmes were successfully completed by Agip, British Petroleum and GERTH. These programmes not only give European oil companies hands on experience of this very important technology but should, in the medium term (5 years), lead to an increase in recoverable reserves. Estimated range from 20-100 million m3 of oil.

A novel pipeline compression station was successfully developed by Borsig; this system has been installed on a test basis on a pipeline and is proving very successful.

- 1978 - 4th Round Projects

Drilling: as reported in 1982, the horizontal drilling technique developed by Gerth has been tested in the heavy oil field of Rospo Mare in Italy. This project constitutes a world pinnacle in this technology.

Production technology: the insert wellhead developed by Shell International Petroleum has been fully tested and reported. This technology should find wide application in the offshore industry. Salzgitter succesfully developed a production system for LNG and associated gas aimed at North Sea conditions; a vital element of this system, i.e. the LNG loading arm, is presently under test at full scale. Gerth projects developed to an advanced stage deep water production equipment and articulated columns. The British Gas fluidised bed reactor is attracting considerable attention.

Enhanced recovery: GERTH carried out successful laboratory studies on solvents for use in tertiary oil recovery. Syminex have examined in a feasibility study the effect of impressed electric current on oil production; this project is now entering a field test stage.

In pipeline technology, GERTH successfully carried out a test on their underwater pipeline repair system "Weldap". Coflexip have successfully developed a cryogenic flexible pipeline which is presently being marketed.

- 1979 - 5th Round Projects

In production technology, Wimpy have developed a system of composite nodes for offshore jackets which considerably reduces weight; this technology should be exploited shortly. C.G. Doris have developed a gravity tower for oil production under North Sea conditions and are presently working with Shell to develop applications suitable for exploiting the Golden Block 31/2 in the Norwegian sector of the North Sea.

In enhanced recovery, Agip have examined methods of producing the Ponte-Dirillo Field in Italy; this project is proceeding satisfactorily and should lead to a positive result in the near future.

In pipeline technology, GERTH have demonstrated the physical feasibility of the J-curve pipelaying method, employing welding by electron beam; this pipelaying method will be necessary for deep water applications.

SNAM and Gerth started development of pipeline repair systems for 1000 m. water depth. Full scale prototype sea trials will be undertaken in the near future.

In storage technology, Single Buoy Moorings demonstrated the feasibility of loading shuttle tankers in 300 m water depths and are presently searching for applications for this technology.

- 1980 - 6th Round Projects

In geophysics, four projects were supported and all have been successfully completed. These projects have led to a greater understanding of the geophysical properties of the earth and their determination through sophisticated seismic techniques.

In production technology, British Petroleum have developed a tanker based production system for exploiting small reservoirs; this system will be implemented towards the middle of 1986 and will have a large impact on Europe's recoverable reserves of hydrocarbons. Salzgitter are developing a system of LNG and LPG production from marginal North Sea reservoirs. Tecnomare are examining further concepts of their already successful steel gravity platform, aimed at a valid concept for 300 m water depth.

In enhanced recovery, projects are under way to examine steam injection, microemulsion injection and carbon dioxide flooding.

In auxiliary ships and submersibles SSOS is developing a closed circuit underwater diesel and a new method of pressure hull manufacturing. Sea trials of full scale prototype are foreseen in the near future. Menck has developed a hydraulically driven underwater pile drive hammer, which is presently used worldwide.

In pipeline technology, GERTH are carrying out a trial to demonstrate at full scale the applicability of their J-curve pipelaying method (see 1979 projects).

In storage technology, Distrigas, in cooperation with Geostock,have carried out a field trial in the North of Belgium to establish the technology of cryogenic storage of LPG and LNG in clay. This project has been successful thus far and has strategic implications for European storage of these hydrocarbons.

- 1981 - 7th Round Projects

In geophysics BGR is carrying out field trials to test magnetotelluric remote reference technique.

In production technology, Shell is developing a homing-in device for blow-out control. Gerth has developed and is testing a riser in resin fibre composites.

Enhanced recovery: Elf Italiana has carried out production tests and comparison between vertical and horizontal production wells. Different pilot plants for injection of steam, surfactants and miscible gas have started.

Miscellaneous. Field investigation into the performance of the foundation piles of Magnus platform is ongoing by BP. An underwater remote controlled diver assistance vehicle, David, developed and tested by ZF Herion, is raising fair interest.

- 1982 - 8th Round Projects

In production technology a down hole pumping system is under development by NEI Peebles. Development of a deep water floating production system (830 m. water depth) by Agip and Tripod tower platform (350 m. water depth) by Heerema are well advanced.

In enhanced recovery a Gerth/Agip combination is running an offshore steam injection pilot plant in the Emeraude reservoir. This is a world first of this kind.

In pipeline technology Arge Supra is building an underwater work and pipeline repair vehicle which will be launched for sea trials shortly.

In transport AEG Telefunken is developing an underwater oil loading system.

Most of the projects of 7th and 8th rounds are still under development.

DIFFUSION OF KNOWLEDGE AND RESULTS

The diffusion of knowledge and results of projects in the Community Hydrocarbons Project Scheme is mainly the prerogative of the contractor. Where results have been positive, they have been presented as papers in technical conferences or have been diffused by means of public advertising.

The Commission has aided the diffusion process by two actions:

a) In April 1979, a Symposium was held in Luxembourg at which the results up to that time, were presented and discussed by Community contractors and other interested oil and gas industry representatives. In December 1984 a second Symposium will be held in Luxembourg again.

b) The Commission regularly holds cooperation meetings. These meetings have a dual purpose: firstly they encourage cooperation between Community contractors working on similar problems and, secondly, they act as a forum for the presentation of results and the discussion of future measures to be taken.

INFORMATION FOR FUTURE PROPONENTS

INVITATION TO TENDER - SUBMISSION OF PROPOSALS

The Commission normally seeks to attract research and development proposals for subvention by the Community by issuing an annual call for tender. Since the provision of subvention may interest a great many proponents, the call for tenders is published in the Official Journal of the Communities.

It has been the practice to publish the call for tenders in July with a closing date for proposals of 30 November.

Research proposals should be submitted in the form required by the call for tender with particular attention being paid to the number of copies required.

Proposals should be sent to:

Directorate for Hydrocarbons
Directorate General for Energy
Commission of the European Communities
rue de la Loi, 200
B - 1049 Brussels

Further information on the programme may also be obtained from the above address.

BREAKDOWN OF PROJECTS BY SECTOR

	Exploration	Drilling	Production	Enhanced Recovery	Transport	Storage
1st round (1975)						
Subsidy granted (in UA)	432 000	9 380 000	13 790 360	1 883 673	11 945 926	1 482 000
Number of projects	(1)	(2)	(6)	(2)	(4)	(1)
2nd round (1976)						
Subsidy granted (in UA)	3 287 933	1 483 200	20 913 651	7 358 101	1 353 500	-
Number of projects	(3)	(1)	(18)	(5)	(3)	(-)
3rd round (1977)						
Subsidy granted (in UA/ EUA)	605 344	2 160 528	29 376 876	3 399 101	5 663 303	290 246
Number of projects	(4)	(1)	(19)	(3)	(4)	(1)
4th round (1978)						
Subsidy granted (in EUA)	1 055 450	4 281 599	19 517 390	4 881 513	5 091 159	1 864 769
Number of projects	(5)	(4)	(21)	(5)	(5)	(3)
5th round (1979)						
Subsidy granted (in EUA)	354 059	-	10 996 650	2 082 395	9 323 705	158 875
Number of projects	(2)	(-)	(10)	(2)	(8)	(1)
6th round (1980)						
Subsidy granted (in EUA)	1 732 299	-	16 434 162	6 331 584	1 749 120	1 849 999
Number of projects	(4)	(-)	(21)	(5)	(5)	(1)
7th round (1981)						
Subsidy granted (in ECU)	301 046	445 895	9 707 578	11 441 194	1 114 161	-
Number of projects	(1)	(1)	(13)	(12)	(6)	(-)
8th round (1982)						
Subsidy granted (in ECU)	95 498	-	14 169 256	7 311 148	4 285 028	2 773 804
Number of projects	(1)	(-)	(21)	(9)	(5)	(3)

GEOPHYSICS AND PROSPECTING

<table>
<tr><td><u>Title:</u>
Telemetry seismic data acquisition system (SN 348)</td><td>Project No.: 1/75</td></tr>
<tr><td><u>Contractor:</u>
SERCEL

<u>Address:</u>
BP 64
F - 44471 Carquefou Cedex

<u>Technical director (or person to contact for further information):</u>
J.C. Naudot</td><td><u>Telephone No.:</u>
(40) 3011 81

<u>Telex:</u>
710695</td></tr>
</table>

Aim of the Project:

Develop and market a seismic data acquisition system to record a large number of channels (up to 480).

This system must avoid the drawbacks of multipair cables, i.e. leakage, cross-talk and industrial noise pick-up.

Project Description:

The project could be divided into three main parts:

- analog data acquisition:

  This part includes low cut and anti-alias filters; it has to handle low level signals (down to .15 uV) with a wide dynamic range (up to 110 db).

  The signals are converted to digital in floating point (15 bit mantissa, 4 bit gain) with a special amplifier-converter.

- digital data transmission:

  After analog to digital conversion the seismic data is transmitted on a two pair cable down to a central unit at a rate of 4 Mbits/s. This data transmission requires special integrated circuits to permit a transmission with a BER better than 10^{-9} although the power drain remains very low (100 mW).

- central unit:

  A minicomputer-based central unit controls the whole data acquisition and performs all the tests. It has to format the data to make it possible to record a geophysical tape with the 6250 bpi GCR standard.

State of Project:

The first field tests were performed December 1976; the first operational system was in field mid 1977, since when the system has been marketed.

Results and applications:

At the beginning of 1984, 107 SN 348 were in operation with over 28 000 field units. The SN 348 is used in every kind of oil seismic exploration apart from deep-water operations in all environments. The ability to record a large number of channels with immunity from cable noise is greatly appreciated by the users.

Title : STUDY OF PARTICULAR PROBLEMS INVOLVED IN SEISMIC PROSPECTION	Project n° : 01.02/76
Contractor : GERTH/CGG Address : 4, av. de Bois Préau 92500 RUEIL-MALMAISON FRANCE Technical director (or person to contact for further information) : Mr Gilbert BLU or Mr Bernard MERCIER	Telephone n° : (1) 749.02.14 ext. 2288 or 2747 Telex : 203 050 F

AIM OF THE PROJECT

The objectives of the project were to test new seismic sources and new arrays of sources and recording devices in order to improve marine seismics.

STATE OF PROJECT

The project started on 1st January 1975 and ended on 30th June 1979.

DESCRIPTION AND RESULTS OF THE PROJECT

This project was carried out jointly by GERTH and CGG.

- Deployment of the Flexichoc source. The apparatus of the multiple flexichoc 506 permitting the deployment of sources is operational.
- Large profile using Flexichoc source. A series of experiments was carried out and the results analysed during 1979.
- Large profile using the vaporchoc source. Testing of the Coflexip hoses has been completed. It appears that this solution requires additional study.
- Geometric control of sources and 'fish'. Tests have been carried out, showing that the system of geometric control of sources and fish is feasible.
- Integral navigation. Tests carried out at the beginning of 1979 were judged unsatisfactory and a re-estimate of the performance one can expect to attain with the system studied showed that it will be difficult to obtain satisfactory performance for navigation.

Title: Research of geophysical methodologies and technologies applicable to particularly complex geological situations	Project No.:01.03/76
Contractor: AGIP SpA Address: P.O. Box 12069, IT - 20120 Milan Technical director (or person to contact for further information): D. Bilgeri	Telephone No.: 02/52 023 227 Telex: 310246 ENI I

Aim of the Project:

To enhance the seismic data quality in complex geological environments using newly recorded data and removing phase distorsions created by the weathering layer.

Project Description:

Accurate recording of new data with a data processing composed to conventional methods. Experimental evaluation of residual static anomalies using a surface consistent technique.

State of Project:

The project has been completed.

Results and applications:

Results of the implementation of the project have been obtained on real and synthetic data and are applied to compute, in source and receiver domains, the correlation between time shifts and phase distortions. Applications consist of a second order correction of residual static values.

Title: Marine seismic source development	Project No.: 01.05/77
Contractor: Horizon Exploration Ltd. (formerly S & A Geophysical Ltd.) Address: Horizon House, Azalea Drive Swanley, UK - Kent BR8 8JR Technical director (or person to contact for further information): Paul Newman	Telephone No.: 0322 68011 Telex: 896050 EXPLOR G

Aim of the Project:

To field test and evaluate the water gun as a seismic source for marine geophysical survey work.

Project Description:

A number of Sodera 80 cubic inch water guns were deployed on the survey vessel Oil Hunter and tested in comparison with conventional air gun sources in various N. Sea areas. Static test measurements were also made. Results were assessed from computer processed results.

State of Project:

Field work was completed during 1977.

Results and applications:

The project demonstrated the viability of water guns as an effective seismic source with several advantages over alternative methods. The industry has been rather slow to exploit these advantages, but is doing so now at an increasing rate. Horizon has acquired more than 35,000 km of marine data using the water gun source and this is being employed by an increasing number of seismic contractors.

Title: Use of transverse waves for seismic prospection	Project No.: 01.09/77
Contractor: Compagnie Générale de Géophysique (CGG) Address: 1, Rue Léon Migaux, B.P. 56 F - 91301 Massy Cédex Technical director (or person to contact for further information): Dominique Michon	Telephone No.: (6) 920 84 08 Telex: 692442

Aim of the Project:

The purpose of this project is to find an automatic and continuous computation to correlate seismic events registered in both P and S modes in order to use S waves for seismic interpretation.

Project Description:

Seismic responses are assumed to be comparable in the P and S modes; in other words, the reflection coefficients are assumed to be of the same order of magnitude in both modes. Since propagation velocity is very different, the seismic responses would appear with a different length in the time scale. Where the P S velocity ratio remains constant, the corresponding parts of P and S seismic traces may be inferred by a simple change of scale. For a window small enough for the γ ratio (Vs/Vp) to be considered a constant in this interval, a scale ratio value would exist for which the correlation between the P wave trace and the S wave trace will be maximum. For each position of the gliding window, the detection of this maximum will give the tie of the two traces and the scale ratio. The computation of the Vs/Vp ratio may be done twice, introducing a redundancy: first, from the progression of the tie position; secondly, from the progression of the scale ratio. A statistical approach is necessary for coloured conventional sections, since the colour code is limited to γ values.

State of Project:

Completed.

Results and applications:

No commercial applications. The original assumption was too limited, the influence of γ variations on the amplitudes is not negligible.

Title: CS 2502 Correlator - Stacker - Demultiplexer	Project No.: 01.11/78
Contractor: SERCEL Address: B.P. 64 F - 44471 Carquefou Cédex Technical director (or person to contact for further information): J.C. Naudot	Telephone No.: (40) 30 11 81 Telex: 710 695

Aim of the Project:

Develop a system for Vibroseis* Oil Exploration, making it possible to acquire in the field the full precision correlated and stacked seismic traces in data acquisition real time.

Project Description:

The main part of the project is a fast processor to correlate in the time domain two signals in real time without any limitation on source signal length.

The correlation process is performed prior to the stack process in order to get more versatility.

The general architecture of the system is modular in order that the needs of the users may easily be met simply by adding processors.

Comprehensive software adapts the sytem to any scheme of field operations.

State of Project:

First field tests have been made in January 1981. From that date the CS 2502 has been marketed and at the beginning of 1984, 50 systems were in operation.

Results and applications:

The CS2502 is very widely used in seismic oil exploration by the Vibroseis* method.

It provides results directly in the recording truck with the same quality as those of processing centres, thus saving a large amount of computer time.

Through the use of the CS 2502, new methods such as simultaneous double source recording could have been promoted.

* CONOCO trade mark

Title : TRANSVERSE SEISMIC WAVES	Project n° : 01.12/78
Contractor : GERTH Address : 4, av. de Bois Préau 92500 RUEIL-MALMAISON FRANCE Technical director (or person to contact for further information) : Mr Gilbert BLU or Mr Bernard MERCIER	Telephone n° : (1) 749.02.14 ext. 2288 or 2747 Telex : 203 050 F

AIM OF THE PROJECT

The aim of the project is to study the lithological and petrophysical evolution of the Dogger carbonated reservoirs (depth about 1 300 metres) and the Rhetian clay-sand reservoirs (depth about 2 200 metres) in the Eastern sector of the Paris Basin, thanks to the variations in the ratio of the compression (P) waves and the transverse (S) waves emitted at the interfaces of these reservoirs.

PROJECT DESCRIPTION

The experiments were carried out at the end of 1979 on Soudron reservoir. 30 kilometres of transverse reflection seismics (S) were recorded, superimposed onto older profiles already recorded in compression waves (P) for conventional exploration. The statigraphic and seismic settings were achieved by several soundings, two of which were the subject of specific measurements using vertical seismic profiles recorded with both types of wave (P) and (S), so as to ensure the best possible respective settings.

P transmission by vibrator - cover 24 - 75 m between traces.

S transmission by the MARTHOR system - cover 24 - 45 m between traces.

STATE OF THE PROJECT

The project was completed at the end of 1983.

RESULTS AND APPLICATIONS

In view of the S wave transmission quality, the study yielded positive results at the relatively shallow Dogger reservoir (about 1 300 m) and through the evolution of the ratio of the S/P speeds, correctly determined the lithological and petrophysical variations encountered during the drillings, thus confirming the hopes invested in the use of transverse waves. On the contrary, in the deeper (2 200 m) Rhetian, the signal to noise ratio (S wave) was too low for the study to be carried to a successful conclusion.

The positive experience achieved with the Dogger reservoir is an encouragement to continue the work so as to extend the field of application of the method to deeper-lying and more "subtle" targets.

Title : SEISMIC SURVEY UNDER SALT DOMES IN MEDITERRANEAN SEA	Project n° : 01.13/78
Contractor : GERTH - CGG Address : 4, av. de Bois Préau 92500 RUEIL-MALMAISON FRANCE Technical director (or person to contact for further information) : Mr Gilbert BLU or Mr Bernard MERCIER	Telephone n° : (1) 749.02.14 ext. 2288 or 2747 Telex : 203 050 F

AIM OF THE PROJECT

The project consisted in developing a seismics method to improve knowledge of the infrasaliferous geological structures in the Mediterranean. By its nature, salt results in complex geometries in the subsoil (from the salt cushion to the piercing dome), preventing conventional seismic theories from being applied in practice (both with respect to acquisition and treatment).

PROJECT DESCRIPTION

The initial stage consisted of acquiring three-dimensional data for reproducing isobath charts ; this could be reinjected into the digital processing system in order to reprocess this data and improve the image of the infrasaliferous horizons.

The methods developed were modelizations that assumed it could be possible in some way to perceive the base of the salt on the unprocessed documents. However, when the method was actually used, it was observed that no coherent energy capable of corresponding to the base of the salt could be exploited.

Despite special dynamic corrections taking into account the radius plot programmes, no improvement in the treatment compared to conventional treatments was observed. Special treatments turned out to be too sensitive to the presumed geometry of the roof of the salt dome.

STATE OF THE PROJECT

The project started on 1st March 1979 and was completed on 30th June 1981.

RESULTS AND APPLICATIONS

This study has proved that it was vain to attempt to apply the laws of geometrical optics, even in a highly sophisticated manner, for excessively disturbed geological forms.

This work has doubtless come too early. At the time, vectorial computers did not exist and one could not envisage the interactive solution of the wave equation on this scale with such accuracy.

These failures have led theorists to completely reconsider the question (the method known as deep migration before addition and the generalized inversion method) ; numerical analysis and data processing tests will probably in the future enable this problem to be re-examined and the data acquired reprocessed.

Title: Development of a procedure for the exploration of areas with poor reflections by the combined application of different geophysical methods, taking the N.W. German Basin as ex.	Project No.: 01.15/79
Contractor: Preussag Ag, Erdöl und Erdgas Address: Arndtstr. 1, D - 3000 Hannover 1 Technical director (or person to contact for further information): Prof Dr G Dohr	Telephone No.: 0511-1232 531 Telex: 922851 preed

Aim of the Project:

Find a concept for an efficient exploration of mineral resources and hydrocarbons in areas with poor quality in reflection seismics.

Project Description:

On the base of the existing results main emphasis was laid on the combined interpretation of results of different geophysical methods. Modern techniques in geophysical data processing were used on a large scale. In this project a new interpretation of the "Gravimetrische Reichsaufnahme" has been carried out, whereas the influence of the sediments down to the Zechstein was approximately eliminated. The results of this interpretation differ from those of the "Reichsaufnahme". This might be caused mainly by the geologic conditions beneath the Zechstein (Lower Permian).

A basin study was made by a synopsis of gravity and aeromagnetic results, indicating the history of the basin. Modelling of aeromatic results as well as interpretation of deep reflection observations in N.W. Germany were used as additional information. More than 60 seismic lines, recorded up to 12 sec. were used.

State of Project:

The project was finished in December 1981.

Results and applications:

Most important information were obtained by "gravity stripping" and additional gravimetric modelling. Hercynian and variscian elements occur in the new gravity map while deep seismic reflection work gives new and qualified information about the lower crust.

The main results derived from gravimetric, magnetic and seismic data is a conception of the figuration of the North-West German Basin which differs from many existing theories. Thus a general way is indicated to explore similar areas which cannot be explored by seismic survey only.

Title: Development of pulse 8/3 range (RHO3) circular positioning system	Project No.: 01.18/79
Contractor: Sea Surveys Address: Kinsale Road Rathmaculing West EI - Cork Technical director (or person to contact for further information): Mr Kavanagh/Mr O'Shea	Telephone No.: (021) 962 600 Telex: 28442

Aim of the Project:
Development of pulses (range RHO/3) circular positioning system.

Project Description:
With the existing chain of pulse/8 stations, positioning cover is insured over the Porcupine Bank considering the first chain of pulse stations, with the development of 2 additional stations in early 1978. However, due to the physical shape of the coastline, there occurs an unfavourable angle of the positioning lines at the point of intersection over the Bank, and the expansion of the pulse/8 pattern caused by the distance from the base line exists.

A solution would be circular ranging, using an "on board" time reference (high stability oscillator) and a sophisticated software package with mini-computer to resolve ambiguities and enable time to be accurately set.

The input from 3 shore stations could be resolved into a circular position line fix. The quality of the fix would be excellent due to wider angles of cut and lack of pattern expansion. This in turn would extend the possible area of coverage.

Results and applications:
The RHO/3 system has successfully proved that accurate circular positioning in a range-range mode could be carried out using an on board time standard. For this, 3 problems were solved:

A. The need to "set-up" before sailing the timing standard, usually an atomic clock drift rate, at a known point in the positioning chain coverage.
B. The inability to monitor accurately the clock drift rate during the voyage.
C. The necessity to return to a known point within the chain coverage, at the end of the voyage, to check the total drift over the period.

Nevertheless, all development was stopped because of the following reasons:

1. Reliable, high capacity, low cyclic time, desk top computers were not available at that time.
2. Development of additional pulse/8 transmitting stations resulting in availability of 6 stations for position fixing, while RHO/3 system could only accept a maximum of 3 ranges only.
3. Requirements appeared for a system which could use a mixture of direct circular ranges and hyperbolic position lines then having higher capacities than RHO/3.

Title: VSP's and surface seismic surveying of a gas reservoir	Project No.: 01.21/80
Contractor: GERTH Address: 4 , ave de Bois Préau F - 92500 Rueil-Malmaison Technical director (or person to contact for further information): Mr Gilbert Blu or Mr Bernard Mercier	Telephone No.: (1)749 02 14 Ext 2288 or 2747 Telex: 203 050 F

Aim of the Project:
To study how vertical and surface seismic profiles may help in the detailed determination of gas reservoir characteristics.

Project Description:
An artificial underground gas storage was selected for experiments, as the geology is very well known because of the large number of wells drilled on the structure.Seismic lines were shot in 1980 in the N.W. part of the gas storage and in 1981/1982 in the S.E. part of the same structure. P wave lines were recorded with a high resolution vibrator using a 32-160 hertz sweep and Soursile, a high weight-drop source. S wave lines were obtained with Marthor and Soursile PS, both of them weight-drop sources generating transverse waves with horizontal polarization. Three vertical seismic profiles were recorded with P and S waves in one well with various reservoir saturations.

State of Project:
The project covered the period 1 March 1980 to 30 June 1982.

Results and applications:
On the P sections the resolving power was of the order of 4-6 m, both with the vibrator and soursile. However, the presence of the gas did not generate any bright spot. This rather surprising fact was attributed to a rather heterogenous diffusion of the gas through the reservoir layer.

Owing to the lack of penetration of the S waves, study of the reservoir by means of a comparison of P and S reflections did not prove possible. A comparison of the spring and autumn P and S reflections did not enable any reflectivity variations to be detected on the reservoir upper limit. It was therefore decided to conduct detailed comparisons of P time interval between one reflector above the reservoir and another one below it. Great care was taken to eliminate all causes of time variation other than that due to the presence or absence of gas. It was found that at places where a movement of gas-water front occurs, the time interval is greater in autumn than during the spring. Systematic, very small deviations were observed, of the order of one millisecond or less. Albeit small, they seem to be meaningful because of their correlation with the gas bubble location.

Title: Further development and test of a new measuring system for offshore refraction seismics for hydrocarbon exploration	Project No.: 01.22/80
Contractor: Institute for Geophysics University of Hamburg Address: Bundesstrasse 55 D - 2000 Hamburg 13 Technical director (or person to contact for further information): Prof. Dr. Jannis Makris	Telephone No.: 040 4123 3969 Telex: 214731 unihhd

Aim of the Project:

For hydrocarbon exploration a new refraction seismic measuring system based on Ocean Bottom Seismographs (OBS) should be developed together with fast data processing and evaluation techniques and tested in a rather complex area offshore western Greece.

Project Description:

In an area offshore W. Greece up to 30 OBSs were deployed every 3 km on 3 profiles of 40 - 60 km length. On the lines 12.5 km shots were fired in regular intervals of 400 m. The analog recorded geophone and hydrophone traces were digitized, filtered, deconvolved and plotted into seismogram sections. The data evaluation was subsequently performed using (1) a combined delay time and ray tracing procedure, (2) a more sophisticated ray tracing technique, deployed interactively and (3) a ray method for computation of amplitudes and synthetic seismograms.

State of Project:

The project was completed in June 1982.

Results and applications:

In the OBS recorded seismic sections the first and later arrivals could clearly be identified up to epicentral distances of 60 km. The sections showed strong wide angle reflections from the cristalline basement. By evaluating these wide angle reflections and the first arrivals the sedimentary structures beyond the top of the above mentioned limestones (acoustic basement in normal incidence reflection seismic sections) down to the cristalline basement in depths of 6-7 km could be deliniated. This shows that additional measurements with the OBS based technique can be a great help in investigating sedimentary structures, where normal incidence reflection seismics fail to penetrate interfaces with high acoustic impedance contrasts.

Title : HIGH RESOLUTION SEISMICS	Project n° :01.23/80
Contractor : GERTH Address : 4, av. de Bois Préau 92500 RUEIL-MALMAISON FRANCE Technical director (or person to contact for further information) : Mr Gilbert BLU or Mr Bernard MERCIER	Telephone n° : (1) 749.02.14 ext. 2288 or 2747 Telex : 203 050 F

AIM OF THE PROJECT

The aim of the project is to develop a low-cost seismics method capable of reproducing geological horizons down to a depth of 1 000 metres, with a resolution in the range of 3 metres.

PROJECT DESCRIPTION

Using surface arrays and appropriate processing sequences, seismic signals are sent comprising a wide frequency spectrum, so as to restore the high frequencies (resolution criterion) and low frequencies (penetration criterion).

The programme of work comprise two seismic campaigns. One on Marienbronn site, characterized by lens-type reservoirs, the other on the Gournay-sur-Aronde gas storage, characterized by reservoirs with a low dip.

STATE OF THE PROJECT

The project started on 1st March 1980 and ended on 30th June 1983.

RESULTS AND APPLICATIONS

The work concerning the Marienbronn campaign did not give the results that were hoped for. This campaign was carried out using explosive charges and a sourcile.

The seismic recordings collected at the Gournay-sur-Aronde site using a vibrator generating non-linear frequency sweeps yielded highly encouraging results. The resolution measured at the well was about 5 metres in layers of the reservoir lying at a depth of 800 metres, with a useful frequency spectrum ranging from 28 Hz to almost 200 Hz. In addition , correlations revealed a posteriori a variation in amplitude of the seismic marker in the maximum gas thickness zone and a shift of the coefficients of reflection before and after injection of gas into the storage volume.

The results of the Gournay campaign have enabled the technique of acquisition of high frequencies in the field to be developed. This must now be furthered by efforts on processing, so that the resolution potential acquired in the terrain is not diluted during the data processing stage.

It is indispensable to continue this work in the field of high resolution so as to guarantee resolutions in the range of 5 to 10 metres in the oil zone (2 000 -3 000 metres).

<table>
<tr><td>Title:
Development of a seismics technology for hydrocarbon prospection in ante-permian coal basins</td><td>Project No.: 01.24/80</td></tr>
<tr><td>Contractor:
GERTH

Address:

4, ave de Bois Préau
F - 92500 Rueil-Malmaison

Technical director (or person to contact for further information):
Mr Gilbert Blu or Mr Bernard Mercier</td><td>Telephone No.:
(1) 749 02 14, ext. 2288
or 2747

Telex:
203 050 F</td></tr>
</table>

Aim of the Project:

Testing of a geophysical method applying the latest advances of technology (high power vibrators and telemetering laboratory) in order to resume prospection for hydrocarbons in the ante-permian coal basins.

Project Description:

4 test lines oriented perpendicular to the major geological accident in the region were recorded along the following profiles:

Profile 1: Saint-Omer region
Profile 2: Arras region
Profile 3: Cambrai region
Profile 4: Valenciennes region.

State of Project:

The project has been completed on 30 June 1981.

Results and applications:

The two Western profiles (1 & 2) were disappointing and yielded no geological data owing to their mediocre quality. On the contrary, the acquisition of the Eastern profiles (3 & 4) was highly encouraging: a geological model was built developing regional knowledge of the structural geology. Profile 4 however showed the limits of conventional processing in complex tectonics.

These results show that the geophysical method used is not universal in nature, being a success in East and a failure in the West. This disparity illustrates the difficulties of seismics in a region with a complicated geology. Additional studies are needed, first to refine the working tool and second, to specify the geological models improving, amongst other things, the scattered disordered reflectors beneath the Midi fault.

Title: Further development and testing of the magnetotelluric remote reference technique	Project No.: 01.27/81
Contractor: Federal Institute for Geosciences and Natural Resources Address: Stilleweg 2, D - 3000 Hannover 51 Technical director (or person to contact for further information): Dr W Losecke	Telephone No.: 0511/6432651 Telex: 923730 bfb(bgr) ha d

Aim of the Project:

The magnetotelluric remote reference technique (RRMT) is to be developed further for routine application to hydrocarbon exploration. An increase in the reliability of the data is obtained with RRMT by reduction of the influence of noise.

Project Description:

The RRMT method uses two precisely synchronised MT equipments placed at two sites A and B: the data of site A are processed together with those of site B for improving the results for site A and vice-versa.

This method is applied for routine use in hydrocarbon exploration. The improvements granted by the RRMT method are tested in such areas as volcanic or Alpine nappes or thick evaporite beds. These activities are accompanied by the development of the corresponding hard and software.

State of Project:

Project started in September 1982 with a duration of 36 months. State of project corresponds to schedule.

Results and applications:

The value for a number of parameters has been established on the basis of measurements and processing results obtained so far. By standardisation of the processing methods, the programme can process the data four times as fast as the earlier versions.

Programmes for calculating the confidence limits were developed for the single and remote reference processing programme package. A filter correction programme for the measuring equipment was integrated in the processing programme package, making temperature corrections possible.

A series of measurements has been made in the Lower Saxony Tectogene and in the Ostholstein region.

Title: New position fixing system for remote sensing of hydrocarbons	Project No.: 01.30/82
Contractor: S.S.L. Address: PO Box 36, Banburry GB - Oxon OX15 JB Technical director (or person to contact for further information):	Telephone No.: 029 573 746 Telex: 24224 Ref 1885

Aim of the project:

The aim of the project is to evaluate the new ratio position fixing system "Navstar" for its suitability as an aid to offshore hydrocarbon exploration. Navstar is a satellite-based system funded by the US Department of Defence. It provides a 24 hour global position to an accuracy of approximately 100m together with velocity and atomic time output.

Project Description:

There are three major phases of the project:

a) selection of receivers and integration into test platforms (aircraft, boat); construction of special purpose data recording systems

b) operational trials
establishment of a test range overland using microwave transponders
conduct flight trials over test range over water
conduct marine test

c) analysis of data; computer printout of tracks and comparisons with other inputs such as mini-ranger Loran C.

State of Project:

The project started on January 1984 with no significant results up to now.

DRILLING

<table>
<tr><td>Title:
Deep sea drilling</td><td>Project No.: 2/75</td></tr>
<tr><td>Contractor:
GERTH

Address:
4, avenue de Bois Préau
F - 92500 Rueil-Malmaison

Technical director (or person to contact for further information):

Mr Gilbert Blu or Mr Bernard Mercier</td><td>Telephone No.:
(1) 749 02 14
ext. 2288 or 2747

Telex:
203050</td></tr>
</table>

Aim of the Project:

The purpose of the study was to investigate problems encountered in drilling in water depths up to 1 000 metres.

Project Description:

At the beginning of the project, the common water depth limits in which drilling was practiced were no more than 200 metres, and the programme of work comprised 3 subjects:

- improving the capacity of existing equipment
- designing and developing new types of equipment
- testing under operational conditions in great depths of water combinations of equipment and drilling techniques.

State of the Project:

The project started on 1 January 1984 and ended on 30 October 1977.

Results and applications:

Conventional mooring is feasible in 1 000 m of water, provided the anchor holding pull can exceed 200 tonnes.

A dynamic positioning system combining a Doppler-sonar with a long base acousting positioning system was studied and tests were pursued within the framework of contract 02.09/77.

The drilling riser was subjected to the following specific studies:

- a computer programme simulating the dynamic behaviour of the riser. The programme was compared with the computer programmes of other companies.
- lightening of the riser with buoyancy materials. Tests were performed on materials supplied by Emerson Cummings and SNPE. Additional work is needed to render this equipment operational.
- emptying the riser following sudden disconnection may endanger it; a theoretical approach was followed by tests performed under contract 02.09/77.
- measurements of fundamental parameters (at the foot of the riser) were made in a prototype pup-joint tested later under contract 02.09/77.

A remote-control system for BOP was built by MATRA and successfully tested on the drilling ships Pelican, Pelerin and Petrel.

Title: The dynamically positioned drillship "Petrel"	Project No.: 3/75
Contractor: Offshore Europe S.p.t. Address: 113, Begijnenvest, B - Antwerpen Technical director (or person to contact for further information): M. Deckers	Telephone No.: 32 31 87 70 Telex: 34129

Aim of the Project:

Operational development of systems and devices in water depths up to 600 m:

(1) Dynamic positioning
(2) Riser and associated equipment
(3) Re-entry and reconnection
(4) Diving and underwater intervention
(5) B.O.P. handling
(6) B.O.P. control
(7) Special drilling equipment

Project Description:

The listed systems and devices which were ultimately incorporated in a newly built d.p. drillship (Petrel), have been subjected to all development stages from early design to actual operation at their nominal capacity: concept design, industrial design, detailed engineering, technical, operational and economic evaluations, ordering, following-up fabrication, integration and tests, installation on drillship, drafting of operating and maintenance manuals, acceptance tests, operation.

State of Project:

The systems and devices have been in operation since May 1976, predominantly in severe conditions and partly in water depths beyond the design capacity. As of 31 May 1984 they have have accumulated 1 760 days of actual operation.

Results and applications:

An early report emphasised the generally good appreciations of the systems performances. The few early age deficiencies have been corrected and it can now be stated that the systems are performing much better than average.

Title: Development of deep water drilling technology	Project No.: 02.06/76
Contractor: Ben Odeco Ltd Address: 29 Bernard Street, UK - Edinburgh EH66 R4 Technical director (or person to contact for further information): J.H. Tolson	Telephone No.: (031) 225 26 22 Telex: 72611

Aim of the Project:
Development of deep water drilling technology.

Project Description:
In 1974 Ben Odeco Ltd (BOL) signed a contract with Scotts Shipbuilding Co Ltd to build a dynamically positioned drillship based on an IHC design for which Scotts held licence rights, but to be developed to drill in 3 000 ft of water. This drillship ("Ben Ocean Lancer") was delivered to BOL in March 1977.

To achieve the drilling system required to operate Ben Ocean Lancer in water depths as deep as 3 000 ft (915 m), BOL has undertaken research and development which has included the following specific studies:

- the study of potential acoustic and associated problems
- the study of marine riser in deep water and the specification of riser flotation and related system
- the design of blow-out preventer (BOP) multiplex controls
- the interfacing of the re-entry systems with dynamic positioning systems.

Results and applications:
A complete study of acoustic problems associated with the various systems employed in the drillship has been undertaken and action taken at all design and construction phases to minimise interference with the acoustic systems. An analysis of data recorded on board the drillship during trials has proved that input and electrical noise are at minimal levels. An acoustic signature of the drillship has been established to act as a guide if problems arise in future.

Marine riser studies resulted in the riser flotation specification now installed on the vessel. The design of a gimballing system for the rotary table and diverter housing was performed. Inital test results were satisfactory. Further monitoring during drilling operations has confirmed the initial findings and has established operating parameters for the specified riser. Blow-out preventer controls compatible with the drillship's design operating capabilities were designed, manufactured and then commissioned aboard the drillship.

In order to achieve the optimum degree of fine manoeuvring during well-head re-entry the Automatic Station Keeping (ASK) system and well-head re-entry side scan sonar/underwater television tool have been interfaced. Successful re-entries have been made at all water depths operated and an external TV facility has been adopted as an additional aid to re-entry.

Title: Deep sea drilling techiques	Project No.: 02.09/77
Contractor: GERTH Address: 4, avenue de Bois Préau F - 92500 Rueil-Malmaison Technical director (or person to contact for further information): Mr gilbert Blu or Mr Bernard Mercier	Telephone No.: (1) 749 02 14 ext 2288 or 2747 Telex: 203050

Aim of the Project:

The aim of this project was to improve the drilling water depth.

State of Project:

The project began on 1 May 1977 and ended on 30 June 1981.

Description and results:

Dynamic positioning

Integration of a Doppler Sonar into the dynamic positioning system, together with improvement to the long-base acoustic positioning system, gave rise to successful trials in November 1979.

Riser

Work carried out covered the main factors that are essential to future drilling operations in very deep waters, namely:

- behaviour of the riser under the action of the longitudinal vibrations encountered at depths of over 1 500 m
- architecture of the riser and its tensioning devices
- decreasing the weight of the riser by using suitable buoyancy materials and building it in new materials such as titanium; fatigue tests were performed successfully on the main parts of a prototype titanium riser but the implementation of this technique is limited by the high cost of titanium; a rapid-actuation connector with rotary rings was designed for this titanium riser. 80 connectors derived from this first connector, for installation on an 18" steel riser, were built. These enabled a 1 700 m prototype riser to be built in a very short time, thus permitting GLP 1 and 2 wells to be successfully drilled in the Mediterranean in depths of water of 1 714 and 1 250 m respectively in 1983
- improvement to the riser foot pup joint, by adding a rod joint sensor and a gas detector. Unfortunately, there was no opportunity to test this equipment in the deep Mediterranean tests.

Re-entry sonar

The re-entry sonar makes it possible to seek out the wellhead on the sea bottom when the ship has no method of accurate location on the surface. A re-entry sonar has been developed and demonstration trials performed.

3 000 m drilling support

Study of a conventional vessel capable of making drillings in 3 000 m water depth was completed end 1980. This vessel will be equipped with dynamic positioning and a derrick with hoisting capacity of 900 t, tensioning devices and heave compensator. The study file drawn up will provide a starting point for study of the naval architecture of such a vessel.

Title : HORIZONTAL DRILLING TECHNOLOGY FOR IMPROVING OIL RECOVERY	Project n° : 02.10/78
Contractor : GERTH Address : 4, av. de Bois Préau 92500 RUEIL-MALMAISON FRANCE Technical director (or person to contact for further information) : Mr Gilbert BLU or Mr Bernard MERCIER	Telephone n° : (1) 749.02.14 ext. 2288 or 2747 Telex : 203 050 F

AIM OF THE PROJECT

Increasing the productivity of wells, the sweep efficiency inside the reservoir and the recovery rate were the aims of the "horizontal drilling technology for enchanced oil recovery" project.

DESCRIPTION AND STATE OF THE PROJECT

The programme of work covered 5 years, from 1st January 1979 to 31st December 1983, comprising amongst other things the execution of an initial 270 metre long horizontal drain at Lacq 90 (1980), a second one 470 metres in length at Lacq 91 (1981), and lastly a third 330 m long drain at Castera Lou (1983), which attained a vertical depth of 3 000 metres.

RESULTS AND APPLICATIONS

The results can be summarized as follows :

- Making particular use of a drilling control technique using continous real-time monitoring of the trajectory by bottom-hole sensors, one can today maintain a horizontal drain several hundred metres in length within a pay zone about 20 metres thick, at least down to vertical depths of 3 000 metres.
- The sampling of cores in horizontal drains with recovery ratios of approaching 100 % is possible and provides information that is highly useful for the geological reconnaissance of distant zones within the reservoir.
- SIMPHOR (instrumentation and measuring system in horizontal wells) provides a neat and low-cost solution to the plotting of electric drilling logs and the perforation of the casing in horizontal wells.
- Likewise, the operations of running down and installing casings and liners to cover long horizontal drains are technically feasible.
- On the other hand, additional work will have to be undertaken to finalize the selective completion techniques in horizontal drains, since not all these drains can be equipped with preperforated liners.
- The excess cost involved in executing and completing a horizontal drain and if necessary exploiting it must be offset by higher productivity and recovery ratios compared to those of a vertical well. It will only be after several years of cumulated production that the final assessment can be made as to the gains in productivity, sweep efficiency and recovery rates.

Title : OIL AND GAS SHOW ANALYSER ON DRILLING SITES	Project n° : 02.11/78
Contractor : GERTH Address : 4, av. de Bois Préau 92500 RUEIL-MALMAISON FRANCE Technical director (or person to contact for further information) : Mr Gilbert BLU or Mr Bernard MERCIER	Telephone n° : (1) 749.02.14 ext. 2288 or 2747 Telex : 203 050 F

AIM OF THE PROJECT

The aim of this project was to develop an instrument that could be used on drilling sites to analyse oil and gas shows present in the cuttings.

PROJECT DESCRIPTION

The following were the objectives of the project:

- to construct a fully-automatic appliance capable of detecting and quantitatively measuring the free HC contents (oil, gas) and those of the hydrocarbons obtained by thermal cracking of the residual organic matter (kerogene),
- to adjust the method of analyzing the results.

STATE OF PROJECT

The project ended in 1980 with the construction of a commercializable site prototype.

RESULTS AND APPLICATIONS

Tests on this instrument at several drilling sites in the Paris and the Aquitaine basins have shown that its design was sound and that it was capable of distinguishing between the hydrocarbons normally present in the rocks, and the accumulation hydrocarbons. Obtained continuously during drilling operations, this data provides information on the movements of the hydrocarbons in the strata crossed. By comparing them with other drilling, one is then in a position to distinguish within a given sedimentary basin the zone most conducive to the accumulation on hydrocarbons.

Title : BUILDING A FIELD-OPERABLE SOURCE ROCK ANALYSER USING PYROLYSIS	Project n° : 02.12/78
Contractor : GERTH - PETROFINA Address : 4, av. de Bois Préau 92500 RUEIL-MALMAISON FRANCE Technical director (or person to contact for further information) : Mr Gilbert BLU or Mr Bernard MERCIER	Telephone n° : (1) 749.02.14 ext. 2288 or 2747 Telex : 203 050 F

AIM OF THE PROJECT

The ROCK EVAL I mother rock analyser, developed jointly by IFP and FINA, was designed principally for manual use in the laboratory. The success encountered by this technique has led to the construction of an automatic instrument for rapid on-site exploitation of the parameters provided by pyrolysis of the organic matter in sedimentary rocks.

PROJECT DESCRIPTION

There were 4 stages to this project :

- design and construction of an appliance for automatic analysis of mother rocks,
- design of a module for analysis of the organic sulphur,
- design of a module for determination of the organic carbon,
- execution of an integrated system grouping a combination of these modules.

STATE OF THE PROJECT

The project started on 1st January 1978 and ended on 31st December 1980.

RESULTS AND APPLICATIONS

The project resulted in the construction and marketing of an automatic mother rock analyser. The bases for analysing the organic sulphur and carbon were established.

Applications for this equipment are of interest to :

- petroleum exploration : characterization of the petroleum potential of the rocks, the quality and the types of mother rocks and the state of evolution of the organic matter (delineation of oil zone)
- tar shales : rapid determination of the pyrolysis oil efficiency
- coals : characterization of the state of evolution and rank.

Title: Optoelectronic cables for submarine use	Project No.: 02.13/78
Contractor: SOURIAU & Co 9-13, rue du Général Galliéni, BP 410 F - 92103 Boulogne Bill. and FILECA Route Nationale n° 1 F - 60730 Ste Geneviève Techical director (or person to contact for further information): J. Bollereau; J.J. Dumont; J. Rollet	Telephone No.: 609 92 00 Telex: 250918 Telephone No.: 422 22 22 Telex: 140300

Aim of Project:

Optoelectronic cables for submarine use.

Project Description:

Feasibility study of an optoelectronic cable capable of

- supplying power to equipment which is fixed or mobile (such as a submarine vehicle) at great depth (15 KVA)
- giving it orders or instructions
- transmitting to the surface images of high definition and at a great rate; transmitting to surface of multiplexed data (3 video images).

State of Project:

Work was carried out at the end of 1982, after the construction of a cable model (length: 100 m), tested up to 1 000 bars.

Results and applications:

The study led to the finalisation of technological methods capable of coping with pressures at very great sea depths:

- electrical continuity, three-faced, 1 500 volts up to 1 000 bars
- optical continuity, three fibers, 50u up to 700 bars
- end connectors (depth and surface).

This feasibility study has given the Souriau and Fileca companies very valuable experience for the study and construction of opto-electronic cables, the need for which will soon be confirmed in oil exploration and exploitation at great depths (1 000 to 3 000 m), metallic nodules (5 000 to 6 000 m) and exploration at greater depths (10 000 m).

PRODUCTION SYSTEMS

Title: Deep sea oil production	Project No.: 4/75
Contractor: GERTH Address: 4, avenue de Bois Préau F - 92500 Rueil-Malmaison Technical director (or person to contact for further information): Mr Gilbert Blu or Mr Bernard Mercier	Telephone No.: (1) 749 02 14 Telex: 203050

Aim of the Project:

The purpose of the project is to study methods enabling oil to be produced in depths of water beyond the 200/300m range and down to 1 000m.

The project was directed towards study of components enabling the production installation to be kept as far as possible on the surface.

State of the Project:

The project started on 1 January 1974 and terminated on 30 June 1977.

Description and results:

Articulated columns: three types of articulated columns were studied:
- columns for the loading station, several were installed in the North Sea
- service columns, one is installed on Frigg N.E. site
- production columns, not yet the subject of any application.

Underwater storage facilities in great water depth installed by means of an articulated column: there have been no applications of this study.

Anchored production supports and the associated production risers: this work covered anchoring systems, the use of drilling semisubmersibles as production support and the design of a discontinuous production riser associated to a subsurface buoy. It has not yet resulted in any applications.

Use of flexible lines for ensuring underwater links. This work has been continued under contract 03.82/79.

Installation and automatic connecting up of flowlines on sea bottom: a test was performed in the Mediterranean in May 1977 in 50m water. This work is being continued under contract 03.35/77.

Study of a TFL tool testing station subsequently built on Pecorade field.

Construction of a test station for studying polyphasic flows comprising a 6" diameter loop of 50m length inclinable to ± 10°.

Construction of an automatic production underwater pilot project on Grondin N.E. fields in a depth of 70m. This pilot station comprises a base equipped with a control unit and three wellheads. Three remote-control systems were tested in succession on this station, which was introduced into service in 1979.

Title: Automated subsea wellhead	Project No.: 10/75
Contractor: Tecnomare SpA Address: S. Marco 2091 I - 30124 Venezia Technical director (or person to contact for further information): Ing. Banzoni	Telephone No.: 708622 Telex: 41484

Aim of the Project:

The design and construction of a prototype automated underwater wellhead.

Project Description:

The remote control system includes, in addition to the electronic sub-system, the subsea electro-hydraulic plant. The system is designed to operate up to 600 mt and at a distance from the terminal platform of 8 km.

State of Project:

The design and construction of a complete remote control system which utilizes three different transmission links (acoustic transmission through the sea water, electrical cable and electrical transmission via flowlines) has been carried out.

Results and applications:

The work has led to the following main results:

a) construction and sea trials of a laboratory of the electronic sub-systems based on the transmission of acoustic signals through the sea water
b) construction of the operative prototype of the electronic sub-system; design and construction of the subsea electro-hydraulic plant and all the auxiliary equipment
c) sea trials of the whole system in 70 mt with transmission both via cable and via acoustic propagation. The positive results of these tests have demonstrated the validity and reliabiilty of the control system.

Title: Acquisition of a production technique for the exploitation of deep sea deposits of hydrocarbons	Project No.: 12/75
Contractor: LEA Address: c/o W.S. Atkins Woodcote Grove, Ashley Road, Epsom UK - Surrey KT1 85BW Technical director (or person to contact for further information): M. Micklethwaite	Telephone No.: 26140 Telex: 23497

Aim of the Project:

Acquisition of a production technique for the exploitation of deep sea deposits of hydrocarbons.

State of Project:

The work on all parts of the LEA study was completed by December 1978.

Results and applications:

The progress on each part of the project was as follows:

1) the study into the sensitivity of steel jackets, lack of knowledge of input properties, has aroused much interest as a means of guiding research both in the general sense and also with respect to specific jacket designs

2) the computer programme 'Asalaunch' has been developed and is being used to check the launching of a number of steel jackets

3) the work on water-structure interaction is being published at several conferences and the techniques developed have been used by other EEC grant holders

4) the design for a jacket for 250 m water depth is complete and lessons learned from it, particularly the method of launch, have been translated to other designs

5) the integrated deck scheme has been offered to several oil companies and is currently being assessed by them.

Title: Development of an automatic mooring system	Project No.: 03.12/76
Contractor: Wharton Engineers Address: Watford Road, Elstree, Boreham Wood UK - Herts WD 63 BT Technical director (or person to contact for further information):	Telephone No.:

Aim of the Project:

Development of an automatic mooring system.

Description and Results:

Phases 1 and 2 have been completed and the following has been established:

1) The requirement of a load sensing equipment with sufficient sensitivity to give the necessary response.

2) The magnitude of the forces acting on a vessel and identification of the rate of change of these forces.

3) Examination of the spring characteristics of various mooring systems at a variety of water depths to assess how much increment was required to change the force on the mooring lines and how quickly it should be carried out.

Phase 3 completed the following tasks:

A. Tank tests have been carried out in conjunction with BP Tanker Co.

B. The traction winch unit has been further developed.

The report on Phase 3 concluded that little further work could be done until a system was sold. Development work has been suspended and a suitable client is being sought.

Title: Exploitation of oil/gas fields in deep waters using floating platforms	Project No.: 03.13/76
Contractor: Tecnomare SpA Address: S. Marco 2091, I - 30123 Venezia Technical director (or person to contact for further information): G. Sebastiani	Telephone No.: 708 622 Telex: 410484 MAREVE I

Aim of the Project:

(1) Definition of the calculation tools and study of the components necessary to design and build floating platforms for industrial applications and (2) Development of a design of a floating platform for exploitation of oil field in deep waters.

Project Description:

After a comparative analysis of different concepts and configurations, a "tension leg" platform was identified as the most suitable. The main data assumed for the design were: (North Sea) water depth: 600 m; payload: 26 000 t. The platform configuration presents the following characteristics: (a) semi-submersible type, with four columns connected by a lattice structure; (b) anchoring system composed of 6 pipes for each column connecting the column to the relevant foundation base; (c) anchoring pipes assembled, by automatic welding, during base lowering; (d) lower and upper ends of pipes rigidly constrained and shaped in such a way as to absorb bending movements.

State of Project:

Completed.

Results and applications:

The platform design has been checked through extensive model basin tests. The proposed anchoring system which eliminates the need for critical components considerably increases the reliability of such types of platforms. Since 1982 AGIP and Tecnomare have started working on a joint research project devoted to the development of the practical technology for the industrial application of the concept in very deep Mediterranean waters.

Title: Intermediate and deep sea production	Project No.:03.20/76
Contractor: VO Offshore Limited (formerly Vickers Offshore Ltd.) Address: Craven House Barrow in furness UK - Cumbria LA14 1AF Technical director (or person to contact for further information): D. le Clue	Telephone No.: 0229 27171 Telex: 65349

Aim of the Project:

To study the tethered buoyant platform concept; develop analytical techniques for evaluating the platform dynamics and to examine the design of specific critical elements of the system.

Project Description:

Early work examined the tethered buoyant platform (TBP) concept and its influence on the other elements of the total system such as risers, manifolds and umbilicals. Small-scale tests were carried out on a simple model to demonstrate the dynamic characteristics of the TBP. The next phase of work included forecasts of offshore production for the North Sea and NW Europe against a background of world energy supply and demand in order to identify the potential demand and most likely applications for innovative engineering solutions such as the TBP.

The comparative economics of alternative development programmes were evaluated and the TBP confirmed as a cost effective producing system for a range of fields and applications. The engineering and analytical developments necessary to demonstrate the feasibility of the concept were defined as a basis for further work. The subsequent design study undertook the development of a suite of analytical techniques for predicting the dynamic loading conditions. In addition, design studies were carried out on specific technical aspects, in particular methods of attaching and adjusting the moorings, riser design and installation techniques.

State of Project:

The project was terminated with the consent of the EEC in 1978.

Results and applications:

The results of the work undertaken demonstrated the feasibility of the concept, but have now been superceded by subsequent developments in this field of technology.

Title: Research and development programme for the production of oil and gas from deep water	Project No.:03.21/76
Contractor: Taylor Woodrow Construction Ltd. Address: 345 Ruislip Road Southall, UK - Middlesex 8B1 2QX Technical director (or person to contact for further information): Mr J.R. Smith	Telephone No.: 01 578 2366 Telex: 24428 TAYWOOD G

Aim of the Project:

To develop systems for the economic exploitation of hydrocarbons from water depths in the range 200 m to 500 m.

Project Description:

The project studies the design, construction, installation and operation of a prestressed concrete articulating buoyant column, called ARCOLPROD, as a drilling, production and storage structure in waters beyond the limits of the continental shelf. Effort is concentrated on its application to Northern North Sea conditions and the ability to construct in identified UK sites.

State of Project:

The project was completed and fully reported in 1980.

Results and applications:

The project culminated in a reference design for 100 000 bpd production in a water depth of 250 m and technical viability of the system was established up to 410 m. The reference design, based upon telescopic construction and deployment, was quality assured by Lloyds register, and received marine approval from London Offshore Consultants. Particular novel features which were extensively researched included a flexible joint based upon an array of synthetic tendons, and the effective completion of structural joints offshore. The complete system was validated in large scale model trials.

Title: Novel offshore production systems	Project No.: 03.27/76
Contractor: The British Petroleum Co. PLC Address: Britannic House, Moor Lane UK - London EC2Y 9BU Technical director (or person to contact for further information): R. Bexon (Technical Director) Mgr COD Div, BTS Dept. (Person to contact)	Telephone No.: 01 920 8000 Telex: London 888811

Aim of the Project:

To develop the concept of a Tethered Buoyant Platform for the production of hydrocarbons from offshore reservoirs in water depths of 150 to 600 m.

Project Description:

The project was proposed as a development programme in 4 phases: Phases 1 and 2 involved preliminary development to confirm essential feasibility, optimisation of key items. Phase 3 would extend detail engineering of the key components for a potential application and Phase 4 would commence manufacturing and testing of the critical new items of equipment and materials.

State of Project:

Phases 1 and 2 were satisfactorily completed. At the conclusion of Phase 2, no application was immediately in prospect to BP, and Phase 3 was modified to cover further work on items of general application only (Phase 3A). With work commencing on Hutton TBP and still no application apparent in BP, Phases 3B and 4 were abandoned.

Results and applications:

Feasibility was established. Computer programmes for platform dynamics and riser behaviour were developed and validated by tank model tests. Both tubular and cable tethers were acceptable.

The results are being used to indicate potential production systems for prospective licence areas and have formed the basis for further work for larger TBPs. The present world oil situation does not favour further large expenditure on the work without a specific application.

Title: Development of a floating natural gas liquefaction plant	Project No.: 03.28/76
Contractor: Consortium 76 (Preussag, Linde, Technigaz) Address: c/o Preussag AG, Leibnizufer 9 D - 3000 Hannover Technical director (or person to contact for further information): Dipl. Ing. D. Meyer-Detring	Telephone No.: (0511) 123 27 32 Telex: 922828

Aim of the Project:

To design a floating natural gas liquefaction plant for the economic utilisation of marginal offshore gas fields.

Project Description:

The floating natural gas liquefaction plant with its components - steel superstructure with process plant and loading system, concrete substructure with LNG storage tanks, feed gas transfer, mooring system - was designed. Model tests were carried out.

State of Project:

The basic design and model tests of hydrodynamic forces and motions in waves as well as tests of liquid motion in LNG tanks are completed. Construction costs were estimated.

Results and applications:

The first phase of an "approval in principle" by a major classification has been obtained.

Title: Production of liquefied natural gas and methanol on platforms	Project No.: 03.29/76
Contractor: Salzgitter AG Address: Postfach 41 11 29 D - 3320 Salzgitter 41 Technical director (or person to contact for further information): R. Holekamp	Telephone No.: 05341-21 3962 Telex: 954 481-0 sg d

Aim of the Project:

In view of the growing importance of offshore oil and gas it was decided to develop a supporting structure (platform) which would be capable to carry a heavy weight gas processing plant, i.e. a facility for natural gas liquefaction or methanol production respectively.

Project Description:

In 1976 there was little knowledge concerning platforms for deckloads of more than 2 500 tonnes. Thus the main task of the project was to develop supporting structures for plant weights of up to 15 000 tonnes. Since the product, LNG or Methanol, had to be intermediately stored offshore, a storage platform also had to be developed. In the course of the study a further problem came up, namely the transfer of the low temperature liquid LNG (- 162°C) from the platform to the tanker.

State of Project:

Time and money available for the project only allowed the development of feasible concepts. Of course, for some components, such as platforms and adapted gas processing plants, basic engineering was carried out.

Results and applications:

In order to obtain a more advanced state, the developers submitted a "second phase" and the work was carried out under Project No. 03.58/78.

Title: Offshore production system "Exboy"	Project No.: 03.31/76
Contractor: Freeman Fox and Partners Address: 25, Victoria Street (Smith Block) UK - London SW1 HOEX Technical director (or person to contact for further information): Dr Browne	Telephone No.: (01) 222 80 50 Telex: 916018

Aim of the Project:

Development of an offshore production system "Exboy".

Project Description:

The project follows up an earlier two-year research and devleopment programme carried out in 1973-1975 on tethered bouyant platforms for use in exploiting undersea resources, particularly in deeper waters. Early work resulted in a design concept covering a wide range of payloads up to 22 000 tonnes and a wide range of water depths. As a basic marine platform, the design can be used or adapted for any purpose connected with offshore oil or gas production.

The project is essentially devoted to the platform structure and its hydrodynamic behaviour, together with its associated anchorage systems, and the present phase takes the development of such a platform to a point where the design is suitable for commercial application.

In 1977 and 1978, work was carried on optimising alternatives for the shape of the floating structure, and assessing their dynamic behaviour under the action of waves, winds and currents. Structural design of the hull and platform was carried out to assess hull weights and characteristics. Tank testing was carried out on a number of basic hydrodynamic forms and various anchorage configurations have been tested in conjunction with different hull shapes. A series of tank tests was carried out in 1978 with the objective of minimising anchorage cable forces, and confirming the behaviour of design derived from computer studies. Experimental results deviated from predicted values by considerable amounts, leading to further tests of models designed to explore various aspects of the behaviour of tethered bouyant platforms under wave action and paying particular attention to the analysis of resonance effects in the anchorage cables.

Title: Mobile platform for power generation based on the gas production of small oil fields	Project No.: 03.32/76
Contractor: Deutsche Babcock & Wilcox Address: Postfach 100 347 48 D - 4200 Oberhausen 1 Technical director (or person to contact for further infromation): E Bitterlich	Telephone No.: (0511) 193 21 Telex: 922851

Aim of the Project:
To prove the economic and technical viability of utilizing marginal gas fields and associated gas for power generating purposes.

Project Description:
For reasons of economic comparison we have taken the Federal Republic of Germany as a significant area of consumption. The price of electric power and primary energy costs in this region are taken as valid for this calculation. Within the framework of this study the conditions of existing marginal fields are investigated and the possible prices for the supply of NG are calculated. Operating time on one location in a marginal field with a production of approx. 300 600 MW is assumed to be 5 to 7 years.

Results and applications:

- Platform concept
  According to the borderline conditions mentioned in the description, we discovered, during the concept phase (ended April 1977), that a floatable jack-up platform carrying the power station would be the most suitable solution. The platform will be set up on a lower structure which has been specially designed according to the prevailing local conditions. The most economic position for the joints was found to be approx. 10m below the surface. Overall dimensions: 66 x 66 x 24 m. Load-carrying capacity for accommodation and equipment: 9 000 t. Proposed location: southern and central North Sea. The special devices of the jack-up system permits the set-up to be made during 70% of the prevailing weather conditions during the summer.
- Power station
  During the conception phase we investigated the following power-generating cycles: gas turbines only; combined cycle with waste-heat boiler; pressurized furnace boiler with gas turbine. The combined cycle with waste-heat boiler produced the lowest generating costs up to 400 km distance from the coast. This holds true for primary energy prices which are within the production costs and return expectations of the oil companies providing the gas.
- Energy transfer
  For technical reasons energy transfer can only be achieved by means of a high voltage direct current system, permitting the transport of electric energy over large distances by subsea cables. The evaluation of already existing cables placed on the surface of the sea bottom revealed their relative sensitivity against damage. Subject of the study is also to evaluate existing trenching methods in view of their performance and to propose possibly corrective action.

Title: Deep sea production systems	Project No.:03.33/77
Contractor: Sir Robert McAlpine & Sons Ltd. Address: 40 Bernard Street, UK - London WC1N 1LG Technical director (or person to contact for further information): M. J. Collard	Telephone No.: 01 837 3377 Telex: 22308

Aim of the Project:

To develop sea bed production systems for use in deep water based on the use of dry one-atmosphere chambers.

Project Description:

Development of systems including full operational design, maintenance study, risk analysis, structure and marine engineering and costing.

State of Project:

Complete.

Results and applications:

Design for a variety of applications in water depths down to 1000 m. Reference cases include a 100 000 bpd field in 500 and 1000 m, and a satellite manifold in 300 m water depth.

Title : N° 2 PRODUCTION PILOT PROJECT IN 100-200 M OF WATER	Project n° : 03.34/77
Contractor : GERTH Address : 4, av. de Bois Préau 92500 RUEIL-MALMAISON FRANCE Technical director (or person to contact for further information) : Mr Gilbert BLU or Mr Bernard MERCIER	Telephone n° : (1) 749.02.14 ext. 2288 or 2747 Telex : 203 050 F

AIM OF THE PROJECT

In order to qualify the relevant technology, the purpose of the project developed under this contract was to test a complete deep water offshore production system under actual operating conditions that was to be installed on Alwyn field, situated in block 3/9 East of the British zone of the North Sea.

PROJECT DESCRIPTION

The refusal by the authorities to allow the associated gas of Alwyn field to be burnt obliged the contracting party to delay development of this field, thus interrupting the associated pilot research project at the study stage.

An alternative solution to continue the present contract has been sought on the Frigg North-East gas field, though resumption of the pilot project on this field was dependent on the agreement of the co-holders of the permit.

STATE OF PROJECT

Since this agreement was not forthcoming, the work was abandoned on 30th June 1980.

RESULTS AND APPLICATIONS

The work performed at Alwyn was confined to study of the overall production scheme and the critical elements.

The work on the subsea well base, studied under this contract, has been used for the design of the subsea template supporting the remote controlled 6 wellheads installed on the North-East Frigg gas-field. This equipment is presently in operation.

Title: Deep-sea production equipment techniques (TEPMP)	Project No.: 03.35/77
Contractor: GERTH Address: 4, avenue de Bois Préau F - 92500 Reuil-Malmaison Technical director (or person to contact for further information): Mr Gilbert Blu or Mr Bernard Mercier	Telephone No.: (1) 749 02 14 ext. 2288 or 2747 Telex: 203050

Aim of the Project:
The purpose of the project was to obtain the engineering data, to build prototypes and to test the critical components for deep sea oil production.

State of the Project:
The project began on 1 March 1977 and ended in September 1981.

Description and Results:

Anchorage: a prototype detector of anchor line faults was tested in the laboratory on samples on anchor cables featuring significant defects.

Riser: the general architecture of the riser was defined and hydrodynamic loads determined on models in a testing tank.

Subsurface link: while not calling the principle of the method into question, experiments conducted in 1978 on a flexible arch in the Mediterranean were a failure, owing to the inadequacy of the logistic facilities and poor sea states.

Laying and connection of a flowline: a method of towing near the bottom was tested in May 1981 in the Mediterranean near St Tropez in a depth of 256m. A 1 000m bundle made of 4" and 2" pipes was installed and both ends were connected to a test wellhead and a test manifold. The whole line was pressure tested to 350 bars and a scraper was pumped from the wellhead to the manifold.

Operations inside wells using TFL technique: TFL successfully adapted to 4" lines, while also enabling the endurance of the equipment to be analysed and the equipment best suited to pumped-tool operations to be selected.

Electrical connectors that can be plugged in underwater: Deutsch and Souriau connectors tested over two-year period, during which several years of underwater operation were successfully simulated.

Horizontal diphasic flow: laminated flow and plug flow conditions tested in 6" loop line. Test equipment and multimetric models developed to enable systematic experiments to be performed (under contract 03/80/79).

Pollution prevention: a simulated programme of the rise of oil and gas through water to the surface was tested and technical manuals on using the design programme were drafted.

Title : MAINTENANCE OF SUBSEA EQUIPMENT EXPERIMENTAL PROGRAMMES ON THE NORTH EAST GRONDIN SUBSEA STATION	Project n° : 03.37/77
Contractor : GERTH Address : 4, av. de Bois Préau 92500 RUEIL-MALMAISON FRANCE Technical director (or person to contact for further information) : Mr Gilbert BLU or Mr Bernard MERCIER	Telephone n° : (1) 749.02.14 ext. 2288 or 2747 Telex : 203 050 F

AIM OF THE PROJECT

The objective of the project was to develop automatic techniques for working on subsea production stations.

PROJECT DESCRIPTION

These "diverless" techniques involving maintenance of subsea wells were tested.

The programme covered the following four main subjects :

Automatic installation and retrieval of guide lines

After connecting an initial guide line with a heave-compensated laying tool, a second tool enables four guide lines to installed simultaneously. They are disconnected by means of an independent device.

Working in the wells by snubbing

This operation was performed with the riser under tension, slides on the riser to compensate for the heave, ball joints at the foot and head of the riser and a fixed snubbing unit on the workover barge. Using this system, it proved possible to acidify a well.

Working outside subsea wells

The main objective was to actuate valves on the wellheads and manifolds by means of an independent handling module hooked on to an observation submarine.

Working outside the wells by a robot

A remote-manipulation robot travelling on rails laid along GRONDIN NORTH EAST station comprising two remote-manipulator arms and one telescopic crane was aimed at making hydraulic, electrical and oil interconnections between the various modules of the station.

STATE OF PROJECT

All the experiments were completed between June 1977 and February 1979.

RESULTS AND APPLICATIONS

Generally speaking, all the experiments were successfully concluded and yielded satisfactory results. In particular, acidification of wells by snubbing has enabled the production of the station to be enhanced.

Title: TFL techniques	Project No.: 03.39/77
Contractor: The British Petroleum Co. PLC Address: Britannic House Moor Lane UK - London EC2Y 9BU Technical director (or person to contact for further information): J.F.B. Marriott	Telephone No.: (01) 920 8000 Telex: 888811

Aim of the Project:

The objectives were to investigate the reliability in general of TFL techniques and equipment, including the degree of control available over long flowlines towards evaluating the use of this technique in the completion of subsea wells in the North Sea.

Project Description:

The project included three stages:

1. An assessment of the performance of a 3" TFL system in a short surface loop at Montrose, Scotland.
2. The development of a method of producing "TFL Quality" welds in offshore flowlines which could readily be applied to the new reel-barge laying methods.
3. The construction of a typical 4" nominal subsea satellite completion at BP's oilfield at Eakring, Nottinghamshire, comprising a 2 000m surface loop and a 440m well, and an investigation of wear, control effects and tool reliability when pumped over long distances for a considerable period.

State of Project:

The project was successfully completed early 1980.

Results and applications:

The tests indicated that TFL techniques could be applied with a considerable degree of success provided sufficient attention is paid to advance preparations and training of personnel.

Title: Novel gas/oil and water/oil separation	Project No.: 03.40/77
Contractor: The British Petroleum Co. PLC Address: Britannic House Moor Lane UK - London EC2Y 9BU Technical director (or person to contact for further information): D.L. Knights	Telephone No.: (01) 920 8000 Telex: 888811

Aim of the Project:

This project aimed at the design, construction and testing of prototype compact gas/oil and water/oil separators following the completion of the early development work funded under EEC contract 04.08/76.

Project Description:

A 60 000 barrel/day compact gas/oil separator based on the use of a cyclone was installed in Kuwait. Extensive tests of the unit showed that target separation efficiency, 5 % volume of gas in the separated oil with no oil carry-over in the gas, was attained with the addition of anti-foaming agent.

The prototype water/oil separator, 12 000 barrels/day, was installed in Qatar. Tests of the separator, which is based on the use of coalescing cartridges protected by a backflushing prefilter, showed that the unit attained the target of less than 0.1 % water in oil and less than 15 ppm oil in water.

Both the gas/oil and water/oil separators met the plan area (less than 25 % conventional system area) and weight (less than 30 % conventional system weight) targets set.

State of Project:

The project was completed on 31 December 1979 and was reported in March 1980.

Results and applications:

A licence to market and sell both systems was granted to a UK manufacturer. So far, 3 compact water/oil units for de-oiling produced water have been sold for North Sea duty.

Title: Water/oil separation by high-speed centrifuge	Project No.: 03.41/77
Contractor: Société Bertin & Cie Address: Allée Gabriel Voisin, BP n° 3 F - 78373 Plaisir Cédex Technical director (or person to contact for further information): René Bourassin	Telephone No.: 16 (3) 056 25 00 Telex: 696231

Aim of the Project:

The project aimed at manufacturing a high-speed centrifuge for separation of water/oil mixture. This equipment will be of reduced size and weight compared with other systems of the same output. The oil content of discharged water will be 10 ppm.

Project Description:

The centrifuge comprises a bowl mounted on an axial shaft bearing a disk stack. A wiper screw conveyor, fastened to the shaft, with a low differential speed with respect to the bowl, trains the solid particles towards a sludge space which is periodically discharged through a hydraulically driven valve. As both speed and flow rate reach high figures, an energy conservation system allows the consumed power to be limited.

State of Project:

The centrifuge is completely designed (calculations and drawings). The prototype is assembled and mounted on the test rig. First mechanical tests have been carried out. A patent has been filed. Alfa-Laval France Co. has placed an option for an exclusive licence grant.

Results and applications:

Low flow rate tests have confirmed the figures predicted for oil/water separation. First tests of the industrial prototype have been satisfactory. The bowl reached its rated speed without any trouble. Progress testing is to be started.

Title : EXTENSION OF USES OF ARTICULATED COLUMNS NEW APPLICATIONS DEEP SEA PROGRAMME	Project n° : 03.42/77
Contractor : GERTH Address : 4, av. de Bois Préau 92500 RUEIL-MALMAISON FRANCE Technical director (or person to contact for further information) : Mr Gilbert BLU or Mr Bernard MERCIER	Telephone n° : (1) 749.02.14 ext. 2288 or 2747 Telex : 203 050 F

AIM OF THE PROJECT

In 1975, three articulated columns have been installed in the North Sea. This type of marine structure appeared very attractive for the production of hydrocarbons after which a research programme was carried out, the aim of which was to find new types of application. Work has mainly dealt with the association of a mono-articulated column to a floating vessel and extension of articulated columns to deeper waters.

PROJECT DESCRIPTION

The present research programme has then allowed to perform :

- the application study of articulated columns in deeper waters, taking into account dynamic phenomena. Adapted computation-tools were developed for this application.
- study of the association to a floating vessel, able to support production facilities and/or to act as storage unit. A general design study, model tests and technological studies led to the consideration of permanent mooring of a semi-submersible or a tanker to a monoarticulated column.
- applications already studied, such as the production column or the workover column and technological tests were developed.

STATE OF THE PROJECT

The project was started on 1st March 1977 and ended on 31st December 1980.

RESULTS AND APPLICATIONS

This research program improved the knowledge and design of articulated columns. Although no application as a permanent mooring device has occurred yet, some results obtained through this contract have been used for industrial application.

Title: Tension leg platform improvement	Project No.: 03.43/77
Contractor: Scott Lithgow Address: Port Glasgow UK - Renfrewshire PA14 5DR Technical director (or person to contact for further information): J. Anderson	Telephone No.: (0475) 42101 Telex: 77192

Aim of the Project:
The 1979 sea test programme had as one of its primary goals, the evaluation of the performance of the TLP and a bank of 12 risers in 600 ft water depth.

Project Description:
The project follows on from an earlier project carried out in 1975 on the one-third scale model TLP which proved that it was a viable concept and that its response in a seaway could be predicted with reasonable accuracy using a linearized frequency domain computer programme. In this earlier project only a single riser was tested.

Results and applications:

This goal was successfully attained with the testing for 95 days of the platform, the risers and auxiliary systems. The tests simulate, by a Froude scaling ration of 1 to 3, the operation of a TLP and riser system at a depth of 600 ft when conducted in 200 ft water depth.

During the programme waves equivalent to 50 ft passed through the structure with no platform or riser malfunction. Particularly noteworthy was the fact that the risers were run successfully during some of the worst sea and weather conditions. The instrumentation and data acquisition system performed well without major problems, and a considerable amount of good, meaningful and viable data was recorded. In addition to the structure, riser, tension member and anchorage data obtained, a great deal of practical knowledge was also gained for example:

a) the towing of the platform to and from the test site locations, with the anchors secured to the bottom of the columns, demonstrated the stability of the TLP under tow at a towing speed much faster than predicted
b) the successful deployment of the anchors from the TLP by the TLP
c) the landing and retrieving of the sea-bed template from the platform confirmed the ability to control the template in a level attitude with a level indicator and subsea TV
d) riser running, latching to mandrel on template and tensioning were successfully accomplished despite tidal current with the aid of TV monitoring at the template
e) deployment of a composite riser bundle with little difficulty, demonstrating the ease of operation from a structure not subject to heave.

The computer programme predicted well the riser response, motion and leg tension response, and platform motion when pulled off location. Thus the effectiveness of the design tools for use on a TLP was confirmed.

Title: Concrete structures for marine protection, storage and transportation of hydrocarbons	Project No.: 03.45/77
Contractor: Dyckerhoff & Widmann AG Address: Postfach 81 02 80 D - 8000 München 81 Technical director (or person to contact for further information): Dr. Finsterwalder	Telephone No.: (089) 925 51 Telex: 05-23036

Aim of the Project:
To obtain further knowledge about concrete structures for production, storage and transport of hydrocarbons.

Project Description:
The project objective was to obtain further knowledge about the following floating concrete structures:
- fixed and floating offshore concrete platforms for production and storage of oil and gas in deeper water and difficult environmental conditions
- floating concrete structures for transport of oil, LPG or LNG
- fixed concrete platform or concrete pontoons for use in arctic conditions.

The project was divided into 4 main phases (see results).

Results and applications:

1. An investigation was carried out on offshore concrete structures with respect to static strength: theoretical and experimental studies were conducted on material behaviour under hydrostatic pressure, shear problem and shell structure problem.

2. Investigation of offshore concrete structures with respect to fatigue strength: when a specimen is exposed to reversible loading tension/compression, presence of water in cracks causes deterioration. Fatigue life of submerged concrete is lower than that of dry concrete.

3. Development of concrete structures for storage and transportation of LNG and oil including thermal effects: normal weight concrete B45 and light weight concrete LB45 were tested. Concrete and steel were tested as to temperature. A test rig with measurement equipment was designed. Tests on structural elements under the influence of temperature and load were performed. Permeability of cracked and uncracked concrete was studied by testing. Design criteria for various offshore structures were issued.

4. Concrete offshore structures subject to ice and arctic conditions: the complex failure behaviour of ice interacting with conical structures has been described. Nevertheless, the results of shear box tests are required for final assessment. Model tests confirmed that ridge conditions, preferably 60° inclined cone, represent the best design condition. The strong influence of the friction coefficient between ice and the surface structure has been shown. The design of the connecting part between cone and platform requires special attention. The coefficient of kinetic friction between concrete and ice must be kept as low as possible.

Title: Riser pipeline installation on existing gravity platforms	Project No.: 03.46/77
Contractor: Shell UK Ltd Address: Shell Mex House The Strand UK - London WC2R 0DX Technical director (or person to contact for further information): W. Visser (UEN 13)	Telephone No.: (01) 257 4000 Telex: 919651

Aim of the Project:

To research the technical and economic feasibility of post-installation of a pipeline riser on an existing offshore gravity structure.

Project Description:

Study the possibility of fixing a pipeline riser to an existing concrete gravity structure.
Analyse the strength of the riser and associated connection system in relation to functional and environmental loads.
Study and compare costs of installing risers by means of a self-supporting tower structure which may be of the articulated type.
Undertake detail design and installation of an 8" riser to an existing concrete gravity structure.

State of Project:

Feasibility studies complete. No further work was considered appropriate with regard to detail design and installation of a pilot riser system.

Results and applications:

The original objective of developing a system of special clamps connected directly to the gravity structure has proved to be unsuitable for riser sections below sea level. The study has also shown that the hitherto design approach for predicting hydrodynamic loads on risers attached to large gravity is inappropriate. Revised design criteria are recommended, based on random wave theory, together with velocity enhancement due to wave diffraction.

Title: Watertight shelters for subsea well heads	Project No.:03.47/77
Contractor: Sea Tank Co Address: 58 A, rue du Dessous des Berges F - 75013 Paris Technical director (or person to contact for further information): M. Vache (C.G. Doris)	Telephone No.: (1) 584 11 64 Telex: 270 263 F

Aim of the Project:

Design concrete shells for the protection, inspection and maintenance of subsea well heads for different conditions of exploitation.

Project Description:

The shell has the form of a multicellular concrete structure with a central corridor surrounded by several independent cells, isolated by floodable chambers. Each cell houses a subsea well head. The central corridor is divided in two levels where pipes and manifolds are housed. A floodable chamber fitted at one end of the corridor allows access to the shelter. The top of each cell is fitted with a watertight steel plug for drilling of well and installation of well head. After completion, it is closed in order to allow further dewatering for human access and works. Concrete has been selected for this project because of its resistance to hydrostatic pressures, corrosion and fatigue.

State of Project:

Studies were abandoned in 1978 and the EEC duly informed.

Results and applications:

None.

Title: Floating natural gas liquefaction plant for offshore liquefaction and loading of associated gas	Project No.: 03.48/77
Contractor: Consortium ALP (Preussag AG, Linde AG, Aker Group Ltd.) Address: c/o Preussag AG, Erdöl und Erdgas Marine Technology Dept. P.O. Box 4829, Bünteweg 2 D - 3000 Hannover 71 Technical director (or person to contact for further information): Dr A. Bath	Telephone No.: (0511) 5105 343 Telex: 175118325

Aim of the Project:

Design of a floating natural gas liquefaction plant for the exploitation of marginal offshore gas deposits in exposed areas.

Project Description:

The system is characterized by the following main components:

- natural gas liquefaction plant, mounted on a floating tension leg anchored platform
- floating LNG storage and offloading platform
- transfer system for LNG and return gas between process platform and storage platform
- feed gas pipeline connecting the LNG process plant to the central production platform.

State of Project:

The basic design of the system has been completed. Price estimates have been made. Model tests have been carried out to investigate the manoeuvrability of an LNG-carrier calling at the floating LNG storage platform.

Results and applications:

The system (Phase I)obtained an "approval in principle" by a major classification society.

Title: Development of an offshore gas gathering system	Project No.: 03.49/77
Contractor: David Brown-Vosper (Offshore) Ltd Address: Graphic House, Castle Street, Porchester UK - Hampshire P016 9PH Technical director (or person to contact for further information): J.A. Guest/E.M. Roger - Smith	Telephone No.: Cosham 83331 Telex: 86156

Aim of the Project:

An investigation of the various natural gas liquefaction processes has been made in this study, enabling a simply operated and safe process to be selected.

Project Description:

Two types of floating process terminal have been considered; a multi-leg semi-submersible unit (MLSS) and a barge-like moored process and storage vessel (MPSV). The MPSV can be constructed in a conventional shipyard and a steel structure is favoured. The MLSS, however, may be constructed of steel or concrete, each having particular advantages. At the initial stage therefore, an investigation of both methods was made for the unit so that the design offering greatest promise could be chosen for detail design.

Results and applications:

Initial investigation of designs for plant capacity at each end of the scale showed that an MPSV with satisfactory seagoing performance would nominally be oversize for a 3.4 MMSCMD plant. The MPSV was found more suitable for larger process volumes and so was chosen for the technical part of the study at a throughput of 10.2 MMSCMD. The MLSS was found to be very sensitive to topside payload and so the production throughput of 3.4 MMSCMD was chosen for this configuration.

A representative site in the northern North Sea was chosen in order to define environmental conditions for each liquefaction terminal. Following definition of the design resuirements, these have been used to develop both schemes in sufficient detail for accurate model scale testing and realistic economic analyses to be carried out. Careful consideration of the requirements of statutory bodies for vessel certification was taken throughout this study, as they have a significant effect upon terminal general arrangements, and every system fitted. In addition to testing physical models of both the MLSS and MPSV terminals, an analogue computer model was developed, initially to parallel the model tests and subsequently to extend consideration of environmental conditions to typical storm rotations of wind and sea.

Operating and capital costs were estimated; after establishing overall costs and system availability, an economic analysis may be made determining potential profitability and sensitivity of profitability to changes in various parameters such as field life, inflation, amortisation period etc.

Title: Improvements to gas detection systems	Project No.: 03.50/77
Contractor: J & S Sieger Ltd Address: 31 Nuffield Estate, Poole UK - Dorset BH17 7RZ Technical director (or person to contact for further information): Mr Denney/Mr Judd	Telephone No.: (02013) 6161 Telex: 41138

Aim of the Project:

To develop improvements to be made to gas detection systems.

Project Description:

Apart from the sudden deterioration in performance of gas detectors due to the presence of poisons, many problems may arise due to the construction and manufacture of such devices.
In practice, catalytic sensors lose activity after prolonged use at high temperatures or after operation in gas rich atmospheres and this may be due to agglomeration of the crystallites of the active component or to breakdown of the refractory support caused either by its inherent chemical instability or by the effects of thermal shock. The latter mode of deactivation is helped by the fact that, using conventional methods of deposition of the refractory metal oxide which involve the thermal decomposition of a water-soluble precursor of that oxide, a poor bond is formed between the metal base and the oxide. It is possible to manufacture a gas detector which displays suitable mechanical strength and resistance to thermal effects by means of chemical vapour deposition of a suitable oxide.

As a result of the present work, it appeared that, although little could be done to prevent the initial interaction of the inhibitor and the catalyst, certain catalyst/support systems recovered very much more rapidly than others from this interaction. One such system was based on tin (IV) oxide.

In an attempt to exploit the ability of tin (IV) oxide to recover from exposure to poisons and incorporate the stability of vapour deposited oxides, preliminary experiments were carried out with a palladium/tin oxide sensor produced by CVD. The device was of similar construction to a pallistor detector and consisted of a helical platinum coil embedded in a bead of tin (IV) oxide. The bead was produced by the vapour phase oxidation of tin (IV) chloride on the heated metal coil and the palladium was applied from a solution containing a suitable salt.

Results and applications:

Results showed that tin (IV) oxide was readily deposited on the coils to procure a thick monolithic coating. After coating with palladium, the resulting bead readily oxidised carbon monoxide and hydrogen and, to a lesser degree, butane. Little catalytic activity was shown in the presence of methane and further work was clearly necessary before a device suitable for trials in an industrial environment could be developed.

Title: Increased production capacity by TFL techniques	Project No.: 03.57/78
Contractor: The British Petroleum Co PLC Address: Britannic House Moor Lane UK - London EC2Y 9BU Technical director (or person to contact for further information): J. F. B. Marriott	Telephone No.: (1) 920 8000 Telex: 888811

Aim of the Project:

Following the successful completion of EEC contract 03.39/77, it was decided to attempt to enhance TFL techniques by making feasible the use of larger diameter flowlines and improved tool application for subsea templates. Contract 03.57/78 was signed to support work in this respect.

Project Description:

The project consisted of the following:

1. The development and testing of a 4" x 2" carrier tool system to allow the use of larger diameter flowlines to satellite wells.
2. The development and testing of a 4" flowline 6-way selector to divert tool strings to any one of several wells in a manifolded subsea completion.

Extensive modifications were made to the Eakring test facility to facilitate the installation and testing of these systems. Comprehensive testing of the carrier tool occupied a period of 6 months; the selector was tested over a period of 3 months.

State of Project:

Testing the carrier tool was successfully completed in March 1981. Testing the selector was delayed due to manufacturing problems but was finally completed by the end of 1981.

Results and applications:

Testing the carrier tool indicated a high degree of reliability for this equipment. The selector design was shown to be generally feasible though improvements in its remote control system were found to be necessary.

Title: Production system for liquefied natural gas and associated gas in the North Sea	Project No.: 03.58/78
Contractor: Salzgitter AG Address: Postfach 41 11 29 D - 3320 Salzgitter 41 Technical director (or person to contact for further information): R. Holekamp	Telephone No.: 05341 21-3962 Telex: 954 481

Aim of the Project:

The main aim of the study was to develop a supporting structure for deeper waters.

Project Description:

The project comprised three main platform versions, all based upon a jack-up type as process plant support. This concept allows the installation of the liquefaction plant in a shipyard. The turn-key facility can be tugged to the location. If the water depth is not more than 60 m, the jack-up platform is used without further auxiliary structures. For a water depth of up to 180 m the jack-up platform is positioned on a substructure. In cases where the water depth goes down to 300 m a tension leg system serves as artificial seabed for the LNG plant carrying jack-up.

State of Project:

Project finished in 1983. The status reached can be described as "detailed engineering", in particular for the TLP system and all offsite facilities, such as floating LNG storage ship, LNG transfer system, LNG tanker mooring and manoeuvring.

Results and applications:

The results were frequently published in technical papers, lectures were given at renowned conferences and technical details were discussed with oil companies. So far there was no chance to realise an offshore LNG production plant. The need for LNG suffered worldwide due to the drop in energy prices and the weakness of leading industrial nations' economy. Nevertheless, the companies involved in the project are convinced that the project finds its intended applications.

Title: Insert wellhead completion system	Project No.: 03.59/78
Contractor: Shell Internationale Petroleum Maatschappij B.V. Address: Carel v. Bylandtlaan 30 NL - 2501 AN Den Haag Technical director (or person to contact for further information): K.W. Brands	Telephone No.: 070-779111 Telex: 31005

Aim of the Project:

Design, manufacture and installation of an insert wellhead completion system offering the inherent safety feature of a protected submudline valve block combined with a low profile above seabed.

Project Description:

The project was conceived in May 1978. Manufacture of the completion equipment and acceptance testing were completed February 1980. Field installation was completed June 1982 by the opening up of the well into the offshore production.

State of Project:

The well has, with the exception of some months to correct a control system problem, been available for production since June 1982.

Results and applications:

To date this is the only insert tree system built and installed.

Title: The design, construction and field testing of a surface controlled subsurface safety valve system in gas and oil wells	Project No.: 03.60/78
Contractor: Shell Internationale Petroleum Maatschappij B.V. Address: P.O. Box 162, NL - 2501 An The Hague Technical director (or person to contact for further information): J.S. Gresham, EP/23.23	Telephone No.: (070) 773056 Telex: 31005

Aim of the Project:

Development of a deepset surface controlled subsurface safety valve system which should lead to improved safety in the production of hydrocarbons through protection of the environment and safeguarding investment.

Project Description:

Following alternative system studies three were selected for further development. However, only two of these, the annulus pressure operated safety valve and the electricsolenoid operated safety valve, have been constructed. Field testing of these two systems is continuing.

State of Project:

The project effectively commenced April 1979 and the field testing will be complete in December 1984.

Results and applications:

The two safety valve systems, which are undergoing field testing in gas producing wells, can be set at any depth in the well and in this respect have a distinct advantage over the more conventional design of hydraulically operated subsurface safety valves.

The electric solenoid safety valve has been undergoing field testing since June 1982 and has proven to be a very reliable system.

The annulus pressure operated safety valve is somewhat complex in design and difficult to reliably control. But testing is continuing.

Title : ARTICULATED COLUMNS PREPARATION FOR INDUSTRIAL REALISATIONS AND NEW DEVELOPMENTS	Project n° : 03.61/78
Contractor : GERTH Address : 4, av. de Bois Préau 92500 RUEIL-MALMAISON FRANCE Technical director (or person to contact for further information) : Mr Gilbert BLU or Mr Bernard MERCIER	Telephone n° : (1) 749.02.14 ext. 2288 or 2747 Telex : 203 050 F

AIM OF THE PROJECT

This contract was a supplement to contract n° 03.42/77. Its aim was to focus on applications of multi-articulated columns and technicological tests.

PROJECT DESCRIPTION

The use of a multi-articulated column for permanent mooring of a converted tanker has been studied. A computer programm has been developed enabling time simulation of the behaviour of the system. It also enabled internal loads in the articulated column to be calculated. A complete series of basin tests was carried out in order to calibrate this computer program.
This theoretical approach was completed by technological studies of important pieces of equipment like jumper hoses or swivel joints. Tests were started on typical seals. For swivel joints tests related to other applications (production column) were also performed. They concerned the checking of bending endurance steel risers and their installation. Six 4" riser-tubing with surrounding equipment were submitted to 16 million cycles which represented a sufficient life-time and gave full confidence in the system.

STATE OF THE PROJECT

The project started on 1st March 1978 and was ended on 30th June 1982.

RESULTS AND APPLICATIONS

Some of these studies and tests had not yet ended at the end of this contract and further developments are being conducted : swivel-joint tests and further studies on mooring systems. Although no application of these studies have yet been made, results are encouraging.

Title : DEEP-SEA PRODUCTION EQUIPMENT (EPMP)	Project n° : 03.63/78
Contractor : GERTH Address : 4, av. de Bois Préau 92500 RUEIL-MALMAISON FRANCE Technical director (or person to contact for further information) : Mr Gilbert BLU or Mr Bernard MERCIER	Telephone n° : (1) 749.02.14 ext. 2288 or 2747 Telex : 203 050 F

AIM OF THE PROJECT

The purpose was to study the main factors common to the most realistic subsea production systems, based on maintaining the greatest possible number of production equipment on the surface.

STATE OF PROJECT

The work was performed from January 1978 until September 1981.

DESCRIPTION AND RESULTS OF THE PROJECT

Riser in composite materials

Tubes of composite material were built with new glass fibres and carbon fibres. Cyclic bending tests performed on pipes of 4" ID - 350 bars service pressure were successfully performed.

Subsea manifold

Study of modular equipment for simplifying the maintenance operations required on the fluid distribution valves of manifolds has resulted in the construction of a prototype module. This module and associated equipment were successfully tested within the framework of contract 03.80/79.

Work inside the wells

Use of extensions in positioning pumped tools enables double completion inside wells to be avoided. The installation of safety valves at considerable depths (40 metres) below the wellhead for single completion has been sucessfully tested, using these extensions.

Safety - Reliability

Systematic analyses by the "faults flowchart" method employed in the nuclear industry has brought about improvement to the production riser, manifold and flowline connection systems. These analyses were highly effective in the flow line connection test preparation.

Combustion of the associated gas during certain purging or emergency operations requires that a method of calculating the effects of thermal radiation be available. A mathematical model of this radiation where no wind is present has been made and tested. Unfortunately, the model studied turned out to be unsuited when the wind factor was taken into consideration.

Flowline connection and laying

A fully automated test for connecting and laying flowlines involving towing of the flowline a few metres above the sea bed has been studied so as subsequently to be able to perform sea trials in deep waters. Sea trials of this method were very successful.

Title: Exploitation of heavy oil	Project No.: 03.67/78
Contractor: British Gas Corporation Address: Research & Development Div. 326 High Holborn, UK - London WC1V 7PT Technical director (or person to contact for further information): Mr K.W S Richards	Telephone No.: 01 242 0789 Telex: 893521/2

Aim of the Project:

To develop a high efficiency process capable of upgrading low value, high sulphur heavy oils to more valuable light hydrocarbons as SNG alone, or SNG plus ethane and light aromatics.

Project Description:

In the process, heavy oils are reacted directly with hydrogen typically at 750°C and at 50 bar in a recirculating fluidised bed reactor, the Fluidised Bed Hydrogenator (FBH). A range of residual oils of SG up to 1.02 have been tested in a 250 mm diameter pilot plant with a throughput of 8500 Nm3/day. Associated studies included recovery of high grade heat from the reactor product stream and the assessment of materials of construction for the process equipment. Scale up of the reactor design was studied on a vessel of 1500 mm diameter using fluidised bed depths of up to 9m. In this vessel nitrogen gas was used at moderate temperature and pressure as the fluidising gas and different designs of recirculating beds were tested.

State of Project:

All of the work has been successfully completed.

Results and applications:

Using atmospheric and vacuum residual oils in the FBH pilot plant, sufficient gasification has been achieved to satisfy the requirements of a fully integrated SNG plant which would have an overall thermal efficiency of about 80%. Separate studies have shown all the process units, including the gasification section, to be technically feasible and practicable. The next stage would be to construct a plant to further develop the FBH reactor and associated systems at or near to commercial scale.

Title : DESIGN OF A HEAVY AND VISCOUS OIL PRODUCTION SYSTEM (ROSPO MARE)	Project n° : 03.68/78
Contractor : GERTH Address : 4, av. de Bois Préau 92500 RUEIL-MALMAISON FRANCE Technical director (or person to contact for further information) : Mr Gilbert BLU or Mr Bernard MERCIER	Telephone n° : (1) 749.02.14 ext. 2288 or 2747 Telex : 203 050 F

AIM OF THE PROJECT

The aim of the project was to design methods and equipment specially adapted to produce and discharge the oil from Rospo Mare field, which is paraffinic, heavy (11 degrees API), has a high sulphur content (6 % by weight) and a high viscosity (2200 cSt at 40°C). Although the asphaltene content is high (17 % by weight), asphaltene deposits appeared low, owing to the naphtenic dominant of the complex.

PROJECT DESCRIPTION

Preliminary study of the pumping facilities eliminated the possibility of centrifugal pumping, the efficiency of which drops too sharply as the oil viscosity rises. Long-stroke pumping transpired to be the most attractive method, since it offered better filling of the bottom hole pump, less fatigue on the rods and took up less area on the surface. A novel measuring bench was built enabling the filling coefficient of the bottom pumps to be accurately evaluated for highly viscous fluids. Long-stroke pumps were tested in this way and it was observed that the downward motion of the piston is difficult to accomplish in the viscous crude, whence the recommendation to use drill collars at the bottom of the well.

Study of flow problems on the surface was centered mainly on transporting the crude in an emulsified form. Laboratory research and tests on a semi-industrial circulation loop showed that by creating a direct emulsion of the oil in water, with an optimum water content of 40 %, one can obtain an emulsion viscosity of about 100 times smaller than that of the anhydrous crude. But there was a lack of stability of the direct emulsion in the event of stopping flow. An emulsifier was selected, delaying the segregation.

Study of the problems of separation, treatment and desalination has shown that indirect emulsions form spontaneously and are stable and relatively difficult to break down : simple decantation adding a chemical is not enough and one has to use an electrostatic installation operating at a temperature of about 100°C.

STATE OF PROJECT

The project ended in December 1983.

RESULTS AND APPLICATIONS

Optimization of the production system has been continued outside the project by building an experimental platform on the Rospo Mare site.

Title: Development programme for composite jacket structures	Project No.: 03.71/79
Contractor: Wimpey Laboratories Limited Address: Beaconsfield Road Hayes UK - Middlesex UB4 OLS Technical director (or person to contact for further information: Dr C J Billington	Telephone No.: (01) 573 7744 Telex.: 935797

Aim of the Project:

To develop the methodology for and test data to support the design of composite offshore structures consisting of piled steel tubular jackets with certain members either partially or totally filled with a cementitious material.

Project Description:

The project is concerned in particular with the fact that the principal design and construction problems associated with tubular steel structures concern the tubular joints, especially in deeper waters where both static and environmental loadings are greater and where there is the further complication of transport limitations on the size and weight of structures.

A preliminary, pre-project feasibility study using limited test data indicated that the use of steel and concrete composite construction could alleviate certain joint design and fabrication problems.

The initial phase of the project was a study to produce outline designs for two structures for water depths of 25m and 140m. In addition to this a comprehensive testing programme was prepared to be performed as the second phase of the project. The testing programme consists of 76 static tests and 14 fatigue tests and is designed to enable parametric equations for both static strength and stress concentration factors (SCFs) to be produced on completion of the work.

State of the Project:

The outline design phase has been completed and the testing programme is approximately 25% complete with the remainder of the work to be carried out in the next two years.

Results and applications:

Initial test results on the static ultimate strength of composite tubular joints indicate that considerable economies might be possible. Work is continuing to establish empirical formulae to account for the effects of composite action on the joint strength and reduction in SCFs.

Title: Deep water gravity tower	Project No.: 03.72/79
Contractor: C.G. Doris Address: 58 A, rue du Dessous des Berges - 75013 Paris Technical director (or person to contact for further information): Mr F. Sedillot	Telephone No.: (1) 584 11 64 Telex: 270263 F

Aim of the Project:

To develop an articulated tower which is designed to support a heavy payload corresponding to an offshore drilling and production system in water depths ranging from 300 to 900 m.

Project Description:

The Deep Water Gravity Tower is mainly composed of a column supported through a laminated rubber articulation by a steel base piled on the sea bed. The column includes a concrete cellular floating, providing the necessary buoyancy, on which the steel deck is set, and a steel jacket working as a rigid tension member. At the lower part of the jacket a ballast chamber is filled with sand and counteracts the buoyancy to provide a permanent compression on the joint.

State of Project:

Studies and tests have been completed at the end of 1982. A second phase of the development study is presently under progress to further investigate some points which are critical or need long duration development.

Results and applications:

The dynamic behaviour analysis and structural design of the tower have been carried out. Possible effects due to the compliant nature of the column have been checked on the equipment. Model tests have been made and have confirmed the dynamic analysis previously made. Most of construction and installation operations have been modelled in wave tank. Extensive fatigue and ageing tests have been performed to assess the feasibility of the laminated rubber, articulated joint and select the most suitable materials. Project has demonstrated the feasibility and large interest of concept.

Title: Cryogenic removal of CO_2 from natural gas	Project No.: 03.73/79
Contractor: Snamprogetti SpA Address: S. Donato, I - Milano Technical director (or person to contact for further information): L. Gazzi	Telephone No.: 5201 Telex: 310246

Aim of the Project:

Construction of a demonstration plant for the cryogenic removal of CO_2 from a natural gas.

Project Description:

The project has been foreseen to have the following phases:

1. Engineering of a demonstration plant for the implementation of the Snamprogetti's "Cryofrac" process for cryogenic CO_2 removal from a natural gas
2. Construction of the plant
3. Tests for the process performances.

State of Project:

The project has been interrupted during the engineering phase due to the unavailability of natural gas from the field owner. A similar project is under consideration based on another gas source.

Results and applications:

Title: Concrete articulated tower CONAT OPP	Project No.: 03.75/79
Contractor: Bilfinger & Berger Bauaktiengesellschaft - Offshore Division Address: Kanalstr. 44, D - 2000 Hamburg 76 Technical director (or person to contact for further information): Heinz Gerhard Butt	Telephone No.: (040) 339 23 123 Telex: 211186

Aim of the Project:

To develop a multi-column articulated production platform for application in the North Sea in deep water, utilising the proved components of the test CONAT.

Project Description:

The integrated deck with its dimensions of 110 x 90 m is supported by four ball joints on four columns which are connected with a gravity foundation by means of ball joints provided with an additional cardan joint in the interior.

State of Project:

The work on the project, financially supported by the EEC, has been concluded successfully on 30 November 1983.

Results and applications:

A 4-column platform for application in the North Sea in 380 m water depth has been developed.

The hydromechanical investigations, the results of which have been verified by tank tests, resulted in an excellent motion behaviour under operational conditions. Since the platform will be constructed vertically, special attention has to be paid to the tow-out procedure. Besides ball joints with hydrostatical bearings, ball joints with statical bearings have been developed.

Title: Systems for recovery of hydrocarbons from small offshore fields	Project No.: 03.78/79
Contractor: Taylor Woodrow Construction Ltd Address: 345 Ruislip Road, Southall UK - Middlesex UB1 2QX Technical director (or person to contact for further information): J R Smith	Telephone No.: (01) 758 23 66 Telex: 24428

Aim of the Project:

To develop a range of floating production systems which can be utilized in the cost-effective exploitation of oil and gas from offshore fields within Continental shelf water depths which are considered to be of marginal economic viability.

Project Description:

The programme examines the role of articulating buoyant columns as manifold and riser support, and as mooring systems for floating production based upon the use of semisubmersibles and tankers. Effort is concentrated on the development and validation of coupled software to describe response in North Sea conditions, the upgrading of systems for more severe environments and cost-effective fabrication and installation techniques.

State of Project:

Technical development and supporting research are complete and the project is in its final reporting stage.

Results and applications:

The outcome of the project has been the development of three reference schemes to conceptual design level, based upon typical North Sea field prospects. Most emphasis has been placed upon a manifold and control column which operates as a field step out. The other systems use the principle of the column with a semisubmersible (for a gas condensate field) and a tanker (for a low GOR oil field).

Title: Development of design techniques for instability problems in offshore structures	Project No.: 03.79/79
Contractor: Taylor Woodrow Construction Ltd. Address: 345 Ruislip Road, Southall UK - Middlesex UB1 2QX Technical director (or person to contact for further information): Mr R D Browne	Telephone No.: 01-578 2366 Telex: 24428 TAYWOOD G

Aim of the Project:

To develop an efficient method for the accurate prediction of structural response and collapse load of concrete cylindrical shells under deep sea loadings. The results were to be used to develop safe and economic design methods, thereby enhancing the potential of concrete structures for the exploitation and storage of offshore hydrocarbons.

Project Description:

The project examines the effects of potentially significant parameters on implosion in an effort to improve the existing knowledge and reduce the areas of uncertainty which at present require the imposition of large safety factors.

State of Project:

The project was terminated in 1983 after completing a literature survey on the subject and carrying out preliminary analytical work.

Results and applications:

From the literature survey, an empirical design method was developed which was shown to provide a lower bound to almost all the available test results.

Title : TESTING OF PRODUCTION TECHNIQUES (E T P)	Project n° :03.80/79
Contractor : GERTH Address : 4, av. de Bois Préau 92500 RUEIL-MALMAISON FRANCE Technical director (or person to contact for further information) : Mr Gilbert BLU or Mr Bernard MERCIER	Telephone n° : (1) 749.02.14 ext. 2288 or 2747 Telex : 203 050 F

AIM OF THE PROJECT

This project, which benefited from the results of earlier work, concerns sea trials of a riser foot manifold and experiments on a 6" pipeline to evaluate the behaviour of links for transferring diphasic fluids (gas and liquid).

STATE OF THE PROJECT

The project began on 1st February 1979 and ended on 31 December 1981.

DESCRIPTION AND RESULTS OF THE PROJECT

Riser foot manifold

The riser foot manifold is a subsea set of valves distributing both the production of hydrocarbons and of fluids required for production (injection gas or water, recirculation or safety fluids). This equipment consists of modular elements that are reraised to the surface for maintenance. A test at sea in shallow depth was prepared to test the automatic positioning and correct connection of a manifold module on a structure simulating the base of a riser foot manifold. The test took place in October 1980 in Pertuis d'Antioche. It was interrupted following breakage of the drilling string used to position the equipment. Since this incident called neither the technique nor the method into question, the test was abandoned at this point so as to have equipment available for the deep water tests.

This deep water test was carried out in conjunction with the flowline laying and connecting test in May 1981 in the Mediterranean, in 256 metres of water, with automatic installation of the elements of the manifold and operating tests at pressure of 350 bars (see also project n° 03.35/77).

Two-phase flow

The interest of this work resides in particular in the possibility of considering long distance conveyance through a single pipeline of the oil and the associated gas produced by marginal deposits.

Experiments using a 6" diameter pipeline on two-phase flow covered the influence of the gradient (- 5 to + 7 °), pressure (15 to 40 bars), temperature and viscosity. Several mathematical models were established to predict the pressure losses encountered when conveying fluids under diphasic conditions.

A computer programme named "PEPITE TRANSPORT" was written and de-bugged on the basis of data collected on oil fields. This programme is operational and is presently used for dimensioning gas and oil pipelines.

Title : SUBSURFACE LINKS FOR DEEP SEA PRODUCTION	Project n° : 03.82/79
Contractor : GERTH Address : 4, av. de Bois Préau 92500 RUEIL-MALMAISON FRANCE Technical director (or person to contact for further information) : Mr Gilbert BLU or Mr Bernard MERCIER	Telephone n° : (1) 749.02.14 ext. 2288 or 2747 Telex : 203 050 F

AIM OF THE PROJECT

This project consisted in testing, in a severe marine environment, the dynamic behaviour of a high pressure hose hung in a catenary near the surface, consisting of an endurance test on a "COFLEXIP' hose with an internal diameter of 240 mm maintained at a high service pressure under the conditions prevailing in the North Sea.

PROJECT DESCRIPTION

The test was performed near FRIGG field, in the Norwegian sector of the North Sea. It took place from September 1982 to August 1983.

The 135 metre long hose was hung between a surface buoy held in position by funicular anchors and a subsurface buoy submerged to a depth of 60 metre and anchored vertically by a gravity anchor. The average span of the arch was 60 metre.

The surface buoy was designed so that its incoercible movements, resulting in dynamic stressing of the hoses, would be similar in a 10 metre amplitude wave to those of 220,000 ton tanker in a 15 metre wave. It consisted of a cylindrical shell 8 metre in diameter and 10 metre tall, equipped at its bottom with a 14 metre disc so as to reduce heave. A hawser pipe inside the buoy limited the curvature at the head of the hose.

41 sensors provided measurements of the behaviour of the hose. They were picked up in the surface buoy and beamed by radio to one of the FRIGG platforms, where they were recorded together with the environmental data.

The high pressure inside the hose was kept throughout the experiment. On completion of the test, the hose was subjected to detailed examination and the mechanical and endurance characteristics of its components compared against those of a new hose. No abnormal damage was detected.

STATE OF THE PROJECT

The work ended on 31 st December 1983 with the examination of the hose tested.

RESULTS AND APPLICATIONS

In view of the considerable number of measurements made, the test has permitted quantitative evaluation of the endurance of a large diameter hose operating at a high service pressure, hung in a catenary near the surface, in a highly severe environment.

Title: Mobile floating platform	Project No.: 03.83/79
Contractor: Deutsche Babcock Aktiengesellschaft Address: Postfach 10 03 47/48 Duisburger Strasse 375 D - 4200 Oberhausen 1 Technical director (or person to contact for further information): Mr Weisemuller/Mr Bitterlich	Telephone No.: (0208) 8331 Telex: 856951

Aim of the Project:
A mobile, floatable platform which, being fixed to the sea bed, when operating can withstand the impact of ice under arctic conditions.

Project Description:
The project comprised several phases:

1. Preinvestigation: to determine the technical problems and the basic information
2. Conceptual designs: it was found that the aim could be achieved by applying ground freezing technology to the purpose. Several designs were produced:
   a) a monocone platform with a conical foot and a central shaft filled with sand, laying on an artificial island to which is applied ground freezing technology
   b) a monocone platform with possibility to disengage the platform from the foundation body. Foundation body lies on a concrete ring or a ring of soil which is frozen
   c) a monocone-type platform with storage facilities inside the platform foundation
   d) a semi-circular breaking body (concrete or steel caissons) on a sand berm, with a movable jack-up platform.

   In the course of the project a cooperation was started with dome petroleum which involved the following activities:
3. Soil investigations about shear and liquefaction of frozen and unfrozen soils
4. Investigation of the use of freezing with caisson retained island
5. Investigation of freezing the dams of the "arctic production and loading atoll" concept.

State of Project:
The project was completed by 31 December 1982.

Results and applications:
To resist the ice pressure, ground freezing technique is a flexible and economical solution to stabilize artificial islands as well as platforms under shallow or deep water.

Dome concepts were examined:

- caisson retained island is stabilized by a frozen ring of soil behind the caissons
- arctic production and loading atoll allows considerable cost savings.

Title: Development of a steel gravity platform for 350 metres water depth	Project No.: 03.74/80
Contractor: Tecnomare SpA Address: S. Marco 2091 I - 30124 Venezia Technical director (or person to contact for further information): P. Gava	Telephone No.: 708 622 Telex: 410484

Aim of the Project:

To develop the design of a steel fixed platform for the exploitation of oil/gas fields in water depths up to 350 m.

Project Description:

Among several configurations considered during the project development, the most suitable one turned out to be a gravity tripod platform. The main substructure components are: three foundation bases; three legs coming from the foundation bases up to the deck; a lower triangular frame; a conductor tower. The platform has been designed for 31m wave height, 30 000t payload and North Sea meteoceanographic conditions. The assembling operations are performed in two phases: assembling in dry dock of the lower part including leg sections of sufficient length and final assembling in sheltered deep waters using floating yards.

State of Project:

Completed.

Results and applications:

The structure configuration and the construction assembling procedures present many advantages such as:

- the main structure is monolithic and critical mechanical connections are avoided
- only proven components and available technologies are used
- it is possible to take full advantage of an extensive sub-assembling of the structural components.

Title: Ultrasonic high accuracy counting for liquid and gas	Project No.: 03.76/80
Contractor: Ultraflux Address: 63, rue du Général de Gaulle F - 78300 Poissy Technical director (or person to contact for further information): J. Pierrat	Telephone No.: 979 26 40 Telex: 696 028 F

Aim of the Project:

Realisation of a high accuracy counting system for liquid and gaseous hydrocarbons using ultrasonic methods.

Project Description:

Study of electronics devices with minimum thermic and ageing shift. Testing of multicord spools of large diameter for liquids and gas.

State of Project:

Final report has been established on 20 January 1983.

Results and applications:

Official approval has to be obtained for the commercial counting. The French official Service des Instruments de Mesures is now proceeding to the necessary examinations and tests.

Title: Single point mooring	Project No.: 03.84/80
Contractor: Tecnomare SpA Address: S. Marco 2091 I - 30124 Venezia Technical director (or person to contact for further information): A. Nista	Telephone No.: 708 622 Telex: 410484

Aim of the Project:

To identify an offshore temporary mooring and loading system for very deep water applications, and also a permanent system. To solve design, construction, installation problems of the identified concepts and to give reliable engineering solutions.

Project Description:

After the identification of the temporary and permanent mooring and loading systems' basic concepts and their preliminary feasibility assessment, the most promising ones were selected and developed for a significant application in 1 000 m water depth.

The temporary system is composed of a compliant structure, fixed on the sea bottom, which supports a rotating superstructure comprising all the necessary equipment such as loading boom, helideck and mooring/loading facilities. For the permanent mooring system, an articulated joke system has been conceived, so as to establish a flexible connection between the storage vessel bow and the mooring tower: it is based on the use of a counterweight to develop the vessel mooring restoring force and to stabilize the system.

State of Project:

Completed.

Results and applications:

The two concepts and design procedures have been precertificated by Det Norske Veritas. As a result of the project, an adequate answer to the development of new concepts/technologies for oil field exploitation in deep water locations has been given. The technical feasibility of the temporary mooring loading system as well as of the articulated joke for permanent mooring have been confirmed by model tests.

Title: Development of a system for the production of methanol from gas offshore	Project No.: 03.87/80
Contractor: Stone & Webster Engineering Ltd. Address: Stone and Webster House, 500 Elder Gate Central Milton Keynes, UK - Milton Keynes MK9 1BA Technical director (or person to contact for further information): Mr E Emerson (Project Manager)	Telephone No.: 668 844 Telex: 826701

Aim of the Project:

Development of an engineering design package to demonstrate the technical feasibility and economic viability of an offshore process plant producing methanol from a gas feedstock.

Project Description:

The scope of the project covers two basic areas:

1. Preparation of a basic engineering package for a 1000 tonne/day offshore methanol plant incorporating the S&W Power Reformer.

2. Preparation of the detail engineering design of a small scale (one tenth full size) plant to demonstrate the operability of the Power Reformer concept.

State of Project:

Engineering design of the full scale offshore plant and of the small scale development unit are well advanced with completion scheduled for the latter part of 1984.

Results and applications:

The development of the engineering package for the full scale plant will essentially demonstrate the technical feasibility of an offshore methanol plant and will also provide basic criteria to enable its application for specific gas sources to be readily assessed. The development of the detail design (and subsequent operation) of a small scale power reformer will enable the operability of this novel concept to be fully demonstrated.

Title: Self-installing satellite production unit anchored in tensioned cables	Project No.: 03.90/80
Contractor: Ateliers et Chantiers de Bretagne Address: Prairie au Duc F - 44040 Nantes Cédex 2 Technical director (or person to contact for further information): C. de Vaulx	Telephone No.: (40) 47 31 32 Telex: 710960

Aim of the Project:

Study of a floating support which would enable functional or complementary units to be connected to a fixed or semi-submersible production platform, either when already in exploitation or during the period of being put in exploration.

Project description:

Two cylindrical piles, connected to the exploitation platform by a tubular lattice boom, support the functional unit. The piles are anchored by taut cables, with a tensioning system - the tension being limited by hydraulic cylinders. The platform is joined by a simple connection, with a quick disconnection device controlled from the satellite itself.

State of Project:

Project completed.

Resulsts and applications:

For this draft project, the application of the "floating satellite" concept has been limited to rather calm seas (Mediterranean sea conditions).

An application in more heavy seas would be technically possible but would lead to a significant increase in the dimensions of the structure, such as the weight as well as the strains applied to the platform. The characteristics of the anchor lines should also be increased.

At present, we consider that over 12 m peak to peak, the dimensions of the structure and of the anchoring system would not be in proportion with the envisaged application. For heavy seas, we shall probably have to come back to a more standard anchor system.

Title: Wellhead monitoring system	Project No.: 03.91/80
Contractor: The British Petroleum Co. PLC Address: Britannic House, Moor Lane, UK - London EC2Y 9BU Technical director (or person to contact for further information): R.J. Smale	Telephone No.: (01) 920 8000 Telex: 888811

Aim of the Project:

Development of a monitoring system for remote subsea oil wellheads.

Project Description:

The project involves the development of a system for monitoring, multiplexing and transmitting data, derived from a number of sensors, mounted on a subsea production wellhead, where diver intervention is not possible.

State of Project:

The work is divided into the following phases:

Phase 1: sensor development and testing

Phase 2: total system development and operational testing.

Work commenced March 1980 and competitive bids were obtained for Phase 1 of the project. A contract was awarded to TRW Ferranti Subsea Ltd, to cover development of:

i) valve position sensor
ii) through flowline train sensor
iii) pressure transducer
iv) temperature transducer.

The development and testing of these four sensors has been successfully completed.

The project is presently under review and consideration is being given as to how best the sensors may be integrated with recent developments in subsea multiplexer technology.

Resulsts and applications:

As part of the project, a survey was carried out by Westcott and Butt (Imperial College, University of London), where the development of a new standard for subsea interfacing is proposed.

Title: Heavy oil platforms: feasibility	Project No.: 03.94/80
Contractor: GERTH Address: 4, avenue de Bois Préau F - 92500 Rueil-Malmaison Technical director (or person to contact for further information): Mr Gilbert Blu or Mr Bernard Mercier	Telephone No.: (1) 749 02 14 Telex: 203050

Aim of the Project:

The overall objective of the project was to study preprocessing of heavy oils at the field, thus enabling them to be conveyed by pipeline over great distances, by lowering their viscosity.

To attain this objective, consideration was given to building an experimental platform comprising the following units: desalination, atmospheric distillation, vacuum distillation, de-asphalting, visbreaking, hydro-visbreaking, a variety of hydroprocessing operations, distillation of the products, storage utilities.

The specific aim of this project was to demonstrate the feasibility of such an experimental platform and to provide the basic engineering.

Project Description:

The project consisted of the following two phases:

1) Feasibility study to define the specifications for the units, to study the characteristics of a number of heavy oils and to compare certain process sequences and the operation of the platform units

   - draughting of the installation layout drawings
   - drafting of the 7 process files
   - execution of an experimental programme for support concerning implementation, processing methods and methods of analysing the heavy oils. In this context, 6 heavy oils were studies in particular: Boscan, Laguna Once, Tia Juana Heavy, Athabasca, Cold Lake and North Battleford.

2) Execution of the basic engineering, comprising a cost estimate and request for tender for the main equipment.

State of Project:

In compliance with the schedule laid down, this project started on 1 February 1980 and ended on 30 September 1982.

Results and applications:

The files established during this project formed the starting point for the work performed subsequently under contract 05.30/81.

Title: Development of columns not sensitive to motion; distillation tests on a motion simulator	Project No.: 03.70/78 03.96/80
Contractor: Linde AG, Werksgruppe TVT München Address: Carl v. Linde Strasse D - 8023 Höllriegelskreuth Technical director (or person to contact for further information): Dir. Dr. W. Baldus	Telephone No.: 7273-0 Telex: 528327

Aim of the Project:

As an alternative to the conventional technology of piping offshore natural gas onshore, a number of studies have been performed investigating the on-site liquefaction or processing of natural gas by floating process plants. Still unsolved was the question of the performance of the columns used for various process operations under sea-induced motion. For the safe and continuous operation of such plants it was absolutely necessary to clarify this problem.

Project Description:

A packed distillation column and a column equipped with a special radial flow tray were tested on a motion simulator. For pilot-scale experiments a very sensitive distillation system (toluene-methyl-cyclohexane) was used to determine the efficiency of both types of column internals under various motion conditions. By using a random motion simulator table (owned by Det Norske Veritas, Oslo), movement in all six degrees of freedom could be produced. The simulator was controlled by computer programmes containing statistical data of sea conditions, combined with the motion response of a floating structure, resulting in a realistic motion of this structure under different sea states.

State of Project:

The project was terminated in July 1982.

Results and applications:

From the tests, the efficiency drop of separation columns on floating structures even under severe weather conditions can be predicted. It has also been shown that a sea-induced motion produces less damage than stationary inclination. The results can be applied for the design of nearly all distillation and washing units mounted on floating platforms. In special cases, boundary conditions such as type of process, type of platform, production field etc. have to be carefully considered.

Title: A sea bed located LNG/LPG production system for marginal offshore fields in deep water areas of the North Sea	Project No.: 03.99/80
Contractor: Salzgitter AG Address: Postfach 41 11 29, D - 3320 Salzgitter 41 Technical director (or person to contact for further information): R. Holekamp	Telephone No.: 05341 21-3962 Telex: 954 481 - 0 sg d

Aim of the Project:

The main objective of this project is to develop a new support system for a small LNG/LPG plant which has the process-technical advantage of fixed platforms and which can be used economically in greater water depths.

Project Description:

The requirements led to the conception of a support system where the process equipment is built within a sea bed located concrete casing. The system consists of a submerged casing with the production unit, the store for LNG/LPG, an articulated tower providing the connection between the casing and the sea surface as well as a transfer system for the products.

State of Project:

The development study will be finished in September 1984.

Results and applications:

This development shall improve the economic exploitation of marginal gas fields in order to guarantee a reliable and constant supply of hydrocarbons.

Title: Development of Single Well Oil Production System	Project No.: 03.101/80
Contractor: The British Petroleum Co. PLC Address: Britannic House Moor Lane UK - London EC2Y 9BU Technical director (or person to contact for further information): P. Heywood	Telephone No.: (01) 920 8000 Telex: 888811

Aim of the Project:

The project involved the development of an itinerant floating production, storage and transportation system for use in the North Sea, but capable of adaption for use in other offshore areas. The special-purpose vessel will be fully dynamically-positioned to remain automatically on location over the subsea wellhead. Wellfluids flow through a rigid riser comprising lengths of standard drill pipe deployed and retrieved from the vessel.

Project Description:

The EEC contract provided support for Phases 1 and 2 which covered the feasibility study and detailed design of the vessel and associated subsea controls. The specifications produced at the conclusion of Phase 2 have been used for tendering purposes and contracts for construction of the first SWOPS system (Phase 3) will be placed during 1984.

State of Project:

A final report on Phase 2 was presented to the EEC in November 1983 which summarises the full scope of studies carried out during detailed design, including reports on parametric studies, model testing, dynamic-positioning simulation studies, machinery optimization to take advantage of produced gas as fuel during part of the operating cycle etc. The resulting hybrid plant of gas turbine and diesel driven generators will be a unique combination for a dynamically-positioned ship under fluctuating load.

The riser handling system has been based on conventional equipment wherever possible but the comparatively shallow water operation (up to 200m) has imposed several restraints on the design of both riser handling system and vessel. Interface between riser and moonpool is of particular concern. To minimise risk of damage and keep the size of moonpool as small as possible a carriage, running on vertical rails, supports the riser head on the centreline of the moonpool. A unique feature of SWOPS is its ability to detach iself safely from the wellhead in seconds should an emergency occur.

Results and applications:

The subsea system has been adapted to allow commingled production from two or more wells, and reference to Single Well has therefore been 'dropped'. The system is now referred to as SWOPS.

Title: Design and development of a fixed steel platform for 650 m. water depth	Project No.: 03.103/81
Contractor: S.S.O.S. Address: Via della Scafa, 19 I - 00054 - Roma-Fiumicino Technical director (or person to contact for further information): Mr Liuzzi	Telephone No.: (06) 645 30 41 Telex: 611156

Aim of the Project:

Design and development of a fixed steel platform for 650 m water depth.

Project Description:

The present fixed platforms can work up to 300 m water depth: feasibility studies to reach oil fields in deeper waters were conducted by several engineering companies and were oriented towards cable tension platforms.

S.S.O.S. project tends on the contrary towards a classic fixed steel platform. The platform has been designed as follows:

- fixed structure on four legs (their bases are anchored on the bottom at 650 m. water depth)
- the legs support jacket and deck structures.

State of Project:

7 different structural solutions were examined: the principal legs divided into 2 modular parts. Strain and vibration analytical tests on each structure permitted us to find the best solution.

Results and applications:

This kind of platform will produce:

- new offshore work never achieved by anyone before;
- encouragement of the development of advanced technologies and the creation of new generation of vessels.

Title: In field development of a Neutrabaric Encapsulation System	Project No.: 03.104/81
Contractor: British Underwater Engineering Ltd. Address: 3rd floor, Trafalgar House Hammersmith Int. Centre, UK - London Technical director (or person to contact for further information): Dr C Baxter	Telephone No.: 01 748 4600 Telex: 928241

Aim of the Project:

To demonstrate the operational benefits of encapsulating subsea production equipment within one atmosphere chambers.

Project Description:

Hydrocarbons Great Britain, as operator of the Morecambe Field in the UKCS, required to obtain reservoir data from a monitoring well drilled into the northern sector of the main gas field.

The wellhead equipment was encapsulated in a wet chamber and working access achieved at one atmosphere to complete hook up and recover data.

State of Project:

One system completely installed and operational.

Results and applications:

The first system has been installed, operational access achieved and data recovered. This project therefore achieved a completely successful result.

Talks are in hand with various operators regarding future deep water applications to encapsulate manifold equipment associated with subsea production systems and for temporary encapsulation of repair sites.

Title: Underwater production system	Project No.: 03.108/81
Contractor: Tecnomare SpA Address: S. Marco 2091, I - 30124 Venezia Technical director (or person to contact for further information): G. Franceschini	Telephone No.: 708622 Telex: 410484 MAREVE I

Aim of the Project:

To study and design an underwater production system for the exploitation of hydrocarbon fields in 1000 m water depth.

Project Description:

The project is subdivided in the following phases:

- definition of the system configuration
- design of the system components, definition of the installation and operative procedures, design of the interfaces with other subsystems necessary for the field exploitation
- design, construction and test of critical component models.

State of Project:

The system general configuration has been defined and the design of the system components is under way.

Results and applications:

The system configuration is based on two subsea parallel and interconnected cylinder containing the production equipment; a monopile supporting on the deck living quarter, power generators and control room. For the cylinder wall a "sandwich" solution has been adopted. A proper computer programme for the analysis of "sandwich" structures has been developed and tested. An acoustic telemetry system for controlling the subsea equipment from the surface is under test.

Title: Modular subsea production system	Project No.: 03.109/81
Contractor: Ateliers et Chantiers de Bretagne Address: Prairie au Duc F - 44040 Nantes Cédex 2 Technical director (or person to contact for further information): J.P. Roblin	Telephone No.: (40) 47 31 32 Telex: 710960

Aim of the Project:

Development of a subsea hydrocarbon production system designed for deep water application. Modular approach for surface maintenance and for diverless guidelineless procedures leads to an original architectural concept.

Project Description:

The project proposes a complete system for hydrocarbon production in deep water: subsea wellheads, flowlines, manifold and riser base manifold are considered as well as connecting and handling tools. The study presents a central manifold gathering the production of eight TFL-serviced satellite wells. Active equipment, subject to possible failures (valves, mechanisms, control equipment), is enclosed in standardized, compact and retrievable wet modules. The project scope of work included feasibility and detailed studies. Some key components - flowline connectors, flowline pulling-in subsea winch - were to be manufactured and tested.

State of the Project:

The project was completed in 1983. A file of detailed drawings was prepared. Tests of key components were successfully carried out.

Results and applications:

The concept was proposed to oil companies. Considered as very attractive for future deep water or marginal field developments, the proposed technology was, however, not deemed economically applicable for present diver-accessible offshore fields.

Title: TFL tool selector development	Project No.: 03.110/81
Contractor: Ateliers et Chantiers de Bretagne Address: Prairie au Duc F - 44040 Nantes Cédex 2 Technical director (or person to contact for further information): J.P. Roblin	Telephone No.: (40) 47 31 32 Telex: 710960

Aim of the project:

Development of a TFL selector (Through Flow Line) designed for use in TFL subsea production. Architectural and environmental aspects compatible with a subsea production system designed by ACB.

Project Description:

The TFL selector is located on a subsea manifold. TFL tools are pumped through a common line from the process unit to the TFL selector which permits the selection of one TFL line among those providing access to each satellite well.

The project scope of work included the feasibility and detailed studies of the TFL selector, the manufacturing of a prototype and a qualification test programme. An 8-way, 4" bore, 5000 PSI working pressure selector was specified. Environmental data and procedures were consistent with the deep water, guidelineless and diverless production system taken as reference.

State of Project:

Project was completed in 1983 by performance of a long duration test focused on the behaviour of the sealing assembly.

Results and applications:

As a result of the tests performed, the unit was found to be satisfactory. Component behaviour was carefully examined and seals were the subject of a specific study concerning disconnections.

The equipment was proposed to oil companies for field developments including TFL service equipment.

Title: The design and development of a guideline-less insert wellhead system	Project No.: 03.111/81
Contractor: Shell Internationale Petroleum Maatschappij BV Address: Carel v. Bylandtlaan 30 NL - 2501 AN Den Haag Technical director(or person to contact for further information): K. W. Brands	Telephone No.: 070 - 779111 Telex: 31005

Aim of the Project:

Design, manufacture and offshore installation of a deep water guidelineless insert wellhead completion system

Project Description:

The project was initiated on 1 December 1981. After the completion of the Phase I design engineering study, it was decided to terminate the project due to an excessive increase in estimated project costs.

State of Project:

Project cancelled.

Results and applications:

Phase I design engineering completed.

Title: Design and development of a Homing-in device for blow-out control	Project No.: 03.112/81
Contractor: Shell Internationale Petroleum Maatschappij B.V. Address: Carel van Bylandtlaan 30 NL - 2501 AN Den Haag Technical director (or person to contact for further information): C.F.M. Heck SIPM EP/22.9 B.C. Lehr KSEPL LRP/5	Telephone No.: 070-771352 & 070-112507 Telex: shell nl 31005 ksepl nl 31527

Aim of the Project:

Design and development of an acoustic homing-in device to be used to determine both the relative distance and direction from a relief well to a blow-out well by monitoring, in the relief well, the noise generated by the blowing well.

Project Description:

During the feasibility study, suitable methods have been established for the measurement and analysis of acoustic signals in a borehole. Then engineering-model logging tools have been designed and constructed to evaluate these acoustic principles under simulated blow-out conditions. During third stage, a prototype homing-in tool is to be constructed, which can be used under actuel field conditions.

State of Project:

The feasibility study, the construction of engineering model tools and drilling of the test holes have been successfully completed. Evaluation of the engineering model tools is in progress. The 3rd stage, the construction of a prototype, is being discussed with the logging industry.

Results and applications:

Passive acoustic methods have been developed. The validity of basic assumptions concerning the sound generated by a blowing well has been confirmed by noise logs recorded in actual relief wells. Engineering model logging tools with associated surface equipment and software have been developed and proved to work according to specifications. The tools are currently being tested in a configuration of three shallow test holes which were specifically drilled for this purpose. Preliminary results demonstrated the validity of the underlying theoretical principles and also proved that both direction to and depth of an artificial sound source can be determined with the help of the engineering model tools. Current measurements concentrate on calculating the accuracy of these measurements and improvement of the same by introducing an acoustic impedance model of the subsurface in the evaluation programme.

Title: Concrete platform for deep seas	Project No.: 03.113/81
Contractor: Sea Tank Co Address: Sea Tank Co/C.G. Doris 58 A, rue du Dessous des Berges F - 75013 Paris Technical director (or person to contact for further information): M. Vaché, C.G. Doris	Telephone No.: (1) 584 11 64 Telex: 270263 F

Aim of the Project:

The aim of the project is to develop a concept based on modular construction and specific assembly methods in order to cope with existing construction site.

Project Description:

The resulting structure looks like a stack of several elementary conventional concrete platforms having 3 or 4 columns each. The elementary structures are built in a conventional way. The major innovation consists of the assembly method between the different elementary structures. Each structure is tilted from vertical to horizontal by differential ballasting. When two elementary structures are horizontal, they are approached and connected. Specific devices and procedures have been developed during the first phase of this study. When the complete structure is achieved, it is towed to side in either the vertical or the horizontal position.

State of Project:

The first phase of the project, the conceptual design phase, was completed in June 1983. A second phase is about to start, the aim being to develop, in more detail, procedures, devices and systems which have been developed during the first phase in order to bring the concept to an industrial level.

Results and applications:

The conceptual design phase has demonstrated the feasibility of the concept from structural, construction/installation and operational view points. Although the project has been presented to most of the major oil companies, there is no application for the immediate future owing to oil and gas overproduction. It is likely that the need to develop deep water fields will arise and our concept is one of the most promising for developmment.

Title: Composite risers	Project No.: 03.115/81
Contractor: GERTH/SNIAS Address: 4, avenue de Bois Préau F - 92500 Rueil-Malmaison Technical director (or person to contact for fruther information): Gilbert Blu or Bernard Mercier	Telephone No.: (1) 749 02 14 ext. 2288 or 2747 Telex: 203050

Aim of the Project:

Offshore petroleum exploitation requires the implementation from the sea bottom to the surface of bundles of extension tubes or "risers". This equipment is all the heavier and bulkier, the greater the depth of water. To reduce their weight, use has been envisaged of pipes in composite materials consisting of resin and glass fibre or carbon fibre.

Project Description:

The work carried out under the present contract follows that concerning contract 03.63/78: "deep sea production equipment", which demonstrated the feasibility of light tubes in composite materials (I.D. 4" service pressure 350 bars), capable of being used in risers (production satellites, water injection lines or mud circulation lines).

Carbon fibre base composite tubes were built first with the same diameter (4 inches), but for a higher service pressure (700 bars), and considered as safety lines for drilling operations in great depths of water and, second, appreciably the same service pressure (300 bars), but with a much greater diameter (12 inches), capable of being used as production discharge lines.

State of Project:

The project started in January 1981 and ended on 30 June 1983.

Results and applications:

The 4-inch 700-bar model tubes showed high crushing strength, but did not reach the safety factor of 3 sought for hydraulic bursting.

As for the 12-inch 300-bar model tubes, these successfully underwent hydrulic bursting, tensile and crushing tests.

In addition, an internal Rilsan lining capable of protecting the tubes that are to carry petroleum effluents, the temperature of which attains 100°C, were studied. As it was not possible to prevent the Rilsan sheath from becoming detached, the work will be resumed on new bases.

Lastly, a polyurethane type outer lining to protect the tubes from possible attack by sea water and to absorb light impacts was tested under severe conditions: it will be used for testing the tubes on the site, at full scale.

Title: Deep water gas production	Project No.: 03.118/81
Contractor: The British Petroleum Co PLC Address: Britannic House Moor Lane UK - London EC2Y 9BU Technical director (or person to contact for further information): Manager, Offshore Division	Telephone No.: (01) 920 8000 Telex: 888811

Aim of the Project:

To develop a production system to enable gas to be extracted from offshore fields in water depths of the order of 350m.

Project Description:

The project has studied alternative field development schemes for two scenarios relating to hydrocarbon production offshore northern Norway, and has then made a more detailed investigation of tension leg platforms suitable for exploitation of the scenarios.

State of Project:

Complete.

Results and applications:

The alternative field development schemes indicated that the tension leg platform would be a technically acceptable and economic concept for hydrocarbon production from the two scenarios. Outline designs for suitable structures have been developed, with emphasis being given to areas of technical uncertainty, including tether systems, installation methods and risers. The designs have been adequate for realistic cost estimates to be obtained. The results have shown that the concept is feasible for the envisaged duties and has indicated where further development is required.

Title: Gas disposal system for deep water	Project No.: 03.119/81
Contractor: The British Petroleum Co. PLC Address: Britannic House Moor Lane, UK - EC27 9BU Technical director (or person to contact for further information): Manager, Offshore Division	Telephone No.: 01 - 920 8000 Telex: 888811

Aim of the Project:

To investigate, define and develop possible gas disposal systems for both emergency and normal production from large offshore gas fields.

Project Description:

The project has studied alternative disposal systems for two scenarios relating to hydrocarbon production offshore Northern Norway. The components for gas disposal systems have been identified and evaluated and then combined into feasible gas disposal system alternatives.

State of Project:

Complete.

Results and applications:

The project examined three alternative disposal systems, namely on-platform, bridge linked or remote and has recommended preferred solutions for the two scenarios. A range of technological problems have been reviewed including blowdown techniques, venting versus flaring, gas dispersion, low temperature difficulties and condensate removal. The applicability of alternative components to the proposed systems has been considered. These include support structures, risers, knock-out drums and flare tips. Areas of general applicability where further development is necessary have been identified.

Title: Downhole pumping system development	Project No.: 03.120/82
Contractor: Peebles Electrical Machines (formerly Parsons Peebles Motors & Generators) Address: East Pilton, UK Edinburgh EH5 2XT Technical director (or person to contact for further information): G. Campbell	Telephone No.: 031-552 6261 Telex: 72125 (PP EDIN G)

Aim of the Project:

To develop and test an electric motor driven modular pumping unit, with a maintenance-free life of at least two years at depths of 3000 m in deep oil wells with a deviation of up to 6°/30 m., at oil temperatures of up to 120° and pressures of 210-350 kg/cm2 even when the motor is full of sour crude oil.

Project Description:

Phase 1. Determination of mechanical and electrical properties of materials; development of pressure/temperature cycling equipment for well-bottom simulation tests; design and manufacture of pressure balance unit; development of winding encapsulation and core corrosion protection systems; investigation of bearing coatings to resist abrasive and chemical effects of sour crude oil containing sand; measurement of dynamic thermal and vibration response of full scale simulated shaft/bearing system; design, manufacture and test of prototype motor to establish dynamic characteristics at full load.
Phase 2. Manufacture of second prototype motor embodying the solutions from Phase 1; life test of motor/pump unit, supplied by static converter equipment, and pumping brine solution contaminated with sharp sand.
Phase 3. Manufacture of a module for testing in a North Sea production well, to be provided by AMOCO. Life-testing in the well environment.

State of Project:

Phase 1 - much of the development work is complete; test apparatus has been designed and commissioned and test programmes are well advanced and are being incorporated into design of Phase 2 prototype motor.

Results and applications:

The dynamic thermal and vibration responses of the shaft/bearing system in the test apparatus have been established with "soft" bearing surfaces. Results so far confirm that bearing temperature rises and shaft dynamic response are satisfactory. Stator core ceramic coating for corrosion protection has been demonstrated to be effective in sea water over the above temperature and pressure range. Some problems with manufacture of balancing of long, slender, hollow shafts have still to be overcome.

Title: Floating production system for deep Mediterranean waters	Project No.: 03.121/82
Contractor: Agip SpA Address: Agip/Tein P.O. Box 12069 I - 20120 Milan Technical director (or person to contact for further information): Mr P Tassani	Telephone No.: (3) 2 - 52027380 Telex: 310246 ENI I

Aim of the Project:

Development of a floating production system for exploitation of oil fields in very deep waters in the Mediterranean sea; reference case the Aquila field.

Project Description:

Analysis and selection of the production system configuration with definition of its main components (platform, risers, tendons, etc.). Development of computer procedure for hydrodynamic and structural analysis. Preliminary design of the floating platform and of the other main subsystems. Definition of construction and installation procedures. Model testing of the platform in ocean basin. Cost of schedule evaluation.

State of Project:

The project is presently in progress and completion estimated at about 60%. The production system architecture has been selected; TLP configuration and main subsystems have been defined. Computer procedures for riser, tendons and platform analysis have been developed. Preliminary design of the platform and of the main subsystems is in course of development as well as construction and installation procedures. Several specific studies (foundation feasibility, hydrodynamic forces, tethers reliability, etc.) have been carried out or are in course of development.

Results and applications:

At the present stage of project progress principal results are the development of new computer programmes or the implementation of existing ones for the analysis of TLP's and risers.

Title: Tripod tower platform	Project No.: 03.122/82
Contractor: Heerema Engineering Service BV Address: P.O. Box 9321, NL - 2300 PH Leiden Technical director (or person to contact for further information): H.W. Dennis/J. Meek	Telephone No.: (071) 31 04 31 Telex: 32483

Aim of the Project:

To design and assess the feasibility of a more economical alternative to the conventional jacket, for a fixed steel platform for use in deep water, typically in excess of 300 m.

Project Description:

Starting from the first principles, a simple geometry was conceived, comprising essentially a single vertical column supported by three inclined legs. These four tubular members were all to be constructed from cold-rolled and welded heavy gauge, high quality, mild steel. The whole structure was to be supported by four individual pre-installed piled foundations.

State of Project:

Following the initiation of the project the feasibility study was broken down into major work areas, namely design, fabrication, floating assembly, tow and installation. During this study a bracing structure was added, both to increase the fatigue life, and to assist in inshore assembly.

Results and applications:

Computer analyses have resulted in the optimum dimensions for a typical depth of water and location. Equipment requirements, material specifications, welding procedures, and fabrication yard layout have been investigated in detail. Model tests have shown the viability of the floating assembly, tow and installation procedures. The concept is now under further investigation for application at several deep water locations, with future study to include laboratory testing of welded steel specimens and grouted connections.

Title: Concrete platforms for Arctic regions	Project No.: 03.123/82
Contractor: Sir Robert McAlpine & Sons Ltd Address: 40 Bernard Str., UK - London WC1N 1LG Technical director (or person to contact for further information): J.A. Derrington	Telephone No.: 01 837 3377 Telex: 22308

Aim of the Project:

To develop concrete gravity designs for production platforms in Arctic regions, based on the experience already gained in the North Sea.

Project Description:

The project is in two phases. Phase I - State of the Art Study - has been to investigate the precise requirements for platforms in the Arctic regions and how to extend the existing design concepts by the application of concrete technology already proven in the North Sea. Phase II will then prepare a detailed design proposal and economic evaluation of one or maybe more particular applications.

State of Project:

Phase I has been completed.

Results and applications:

Study of the exploration programme and forecasts, environmental conditions and the properties followed by discussions with operators and government agencies to determine requirements. Design experience has been used to formulate outline designs for platforms for oil or gas production in the Beaufort Sea and the Scotia shelf and Grand Banks area.

Title: Development of a deep water barge	Project No.: 03.124/82
Contractor: Taylor Woodrow Energy Ltd. Address: Taywood House, 345 Ruislip Rd Southall, UK - Middlesex UB1 2QX Technical director (or person to contact for further information): Mr J P Gibson	Telephone No.: 01 578 2366 Telex: 24428

Aim of the Project:

To carry out, and report on, the development of a Barge Mounted Production and Storage System (BPSS). In particular to prove that the system is a commercial viable contender for the exploitation of oil fields in the North Sea, with water depths from 100 m to 300 m and greater.

Project Description:

The project has been divided into three phases.

Phase 1: The computer model of the Barge hydrodynamics was confirmed by model tests. Additionally an outline design for a BPSS in 300 m of water was produced and costed.

Phase 2: A BPSS for 100 m of water has been designed and detailed costing and economic analyses are being carried out. Lloyds "approval in principle" is to be obtained.

Phase 3: The areas of uncertainty disclosed by Phase 2 will be studied. Additionally, in conjunction with an oil company, an engineering design will be produced for a specific application.

State of Project:

Phase 2 will be completed by the end of July 1984.

Results and applications:

The BPSS will have better heave motions than existing semi-submersibles, will cost less than a semi-submersible production system with in-line storage. With storage it is similar in cost to a semi-submersible based system without in-line storage. Equipment used for BPSS can all be of field proved design. It is commercially and technically attractive in water depths greater than 200 m for oil fields with large quantities of associated gas. It lowers the threshold of field viability and provides substantially greater deck load capacity than a semi-submersible based system.

Title: Single Well Oil Production System (SWOPS): sea trials of SWOPS re-entry equipment	Project No.: 03.126/82
Contractor: The British Petroleum Co. PLC Address: Britannic House, Moor Lane, UK - London EC2Y 9BU Technical director (or person to contact for further information): R.C. Toft	Telephone No.: (01) 920 8000 Telex: 888811

Aim of the Project:

Preliminary design of the SWOPS vessel (see project 03.101/80) had indicated the magnitude of riser angle and tension and associated operational procedures that could be expected in routine and emergency situations. This project sought, by both onshore and offshore testing, to prove the capability of the prototype SWOPS subsea re-entry equipment to meet those conditions.

Project Description:

The project was divided into two parts: stage one using a purpose-built test structure, and stage two using a semi-submersible drilling rig and dummy wellhead in the central North Sea area. Stage one was conducted onshore using the prototype re-entry equipment and loads were applied by means of levers and hydraulic cylinders. Stage two utilised purpose-built components intended to simulate the SWOPS riser design and was conducted offshore on a suspended drilling wellhead.

State of Project:

Both parts of the project are now complete, a small amount of development work has been identified and is to be carried out as an extension of this project.

Results and applications:

The testing programme has indicated some undesirable contact at the mating faces. Further design of the SWOPS concept has indicated that the number of control ports at the interface should be increased. New components are to be manufactured and tested onshore, taking into account these aspects, and the resulting assemblies shall form the basis of the equipment to be used on the SWOPS project.

Title: Deep water gravity tower	Project No.: 03.127/82
Contractor: C.G. Doris Address: 58A, rue du Dessous des Berges F - 75013 Paris Technical director (or person to contact for further information): F. Sedillot (C.G. Doris); L. Des Deserts	Telephone No.: (1) 584 11 64 Telex: 270263

Aim of the project:

Within a previous separate phase, the concept of the deep water gravity tower had been developed. The purpose of this present project is now to further investigate some particular points related to this articulated structure, in order to reach a sufficient level of technological development.

Project Description:

The main items studied are the following:

- detailed dynamic response, fatigue and seismic analysis. This study includes the assessment of stress history corresponding to wave and wind spectra, and steel node finite element analysis to establish exact stress concentration factors
- study of drilling and conductor systems, and risers
- tests on the laminated rubber articulated joint, including compression-shear fatigue tests, laboratory and deep sea environmental tests
- detailed construction and installation procedures.

State of Project:

At the present date, September 1984, the first two items listed above are almost completed and the other ones are still under progress. A complete report should be issued early in 1985.

Results and applications:

- A new methodology for fatigue analysis has been established. It is based on the particular geometry of the structure which allowed a constant relationship between forces in members arriving at a node, for a given wave heading. This methodology permitted an optimization of steel quantities.
- Detailed fatigue analyses have been made on conductors and risers, and their installation and connection procedure has been developed.
- Extensive laminated rubber joint tests are under progress. Although it is too early to draw definite conclusions, those tests seem to confirm the very high characteristics which can be obtained with these materials.
- Detailed construction and installation procedures have been prepared, including marine operations and equipment specification.

Title : SEA TRIALS OF COMPOSITE TUBES	Project n° : 03.128/82
Contractor : GERTH - SNIAS Address : 4, av. de Bois Préau 92500 RUEIL-MALMAISON FRANCE Technical director (or person to contact for further information) : Mr Gilbert BLU or Mr Bernard MERCIER	Telephone n° : (1) 749.02.14 ext. 2288 or 2747 Telex : 203 050 F

AIM OF THE PROJECT

The results of the work carried out under contracts 03.63/78 and 03.115/81 concern the development of high performance composite tubes capable of reducing the weight of the vertical links between the bottom and surface of offshore petroleum exploitation systems. They have been judged sufficiently interesting to justify considering making a sea trial to validate the use of these tubes as "peripheral lines" of drilling risers. This test was conceived so as to simulate on full scale and for a significant period the behaviour of the tubes thus applied, from the standpoint of both installation and endurance.

PROJECT DESCRIPTION

Three stages of work were carried out :

- preparation of the trial, comprising definition of the testing conditions, dimensioning of the "booster" type composite tubes (4 inches - 350 bars) and "kill and choke" type tubes (4 inches - 700 bars) and verification of these dimensions on short model tubes manufactured for this purpose
- procurement of the experimental elements, comprising setting up the manufacturing facilities, and producing and testing these elements. Two control tubes of each type were first manufactured and tested, under hydraulic bursting pressure and also under buckling at service pressure
- sea trials proper on two composite lines each consisting of two 15 metre long tubes, one of the "booster" type, the other of the "kill and choke" type ; these two lines were located just below the telescopic joint of the drilling riser of platform P 84 in the North Sea operation, being subjected to the mechanical stresses caused by the waves and current, and to the physical and chemical constraints of the marine atmosphere and solar radiation.

STATE OF THE PROJECT

The sea trial phase is now complete and the tubes are under expertise.

RESULTS AND APPLICATIONS

During the trial, which was carried out through three consecutive drilling periods (in all 136 days), the composite lines were regularly raised to service pressure (22 times in all). The "kill and choke" tubes all behaved satisfactorily throughout the trial ; this was not the case for the "booster" tubes, which proved themselves sensitive to transport and handling conditions (punching or crushing).

Title: Gas offshore Northern Norway Development projects "Subsea production systems"	Project No.: 03.129/82
Contractor: The British Petroleum Co. PLC Address: Britannic House, Moor Lane UK - London EC2Y 9BU Technical director (or person to contact for further information): Manager PLT Div., BTS Dept.	Telephone No.: 01 920 8000 Telex: London 888811

Aim of the Project:

To develop and design subsea production systems for the exploitation of hydrocarbon resources in 350 metres water depth in the Northern North Sea, 150 km from shore.

Project Description:

The project involved a phased investigation of alternative subsea field development methods and the associated systems and equipment required to develop a large gas or oil field. Engineering was performed to establish feasibility, identifying and developing new necessary equipment. A modular approach was pursued and the aspects of diverless installation, access, maintenance, repair and operating techniques were developed. Field production, development schedules and cost estimates were prepared.

State of Project:

First two phases have been satisfactorily concluded. Final prototype equipment manufacturing and testing phase of work have been deferred due to present market and tax conditions.

Results and applications:

The results of the study indicate that it is feasible to engineer a modular subsea production system for offshore Northern Norway. Emphasis should be placed on further optimisation of the basic field development schemes, component reliability and rigorous testing. Essential components and modules identified by the study should be manufactured and tested. A minimum sized subsea facility should be built to enable critical integrated land test programmes to be carried out.

Title: Design of a floating production facility for use on marginal fields	Project No.: 03.131/82
Contractor: Britoil PLC Address: 150 St Vincent Str. UK - Glasgow G2 5LJ Technical director (or person to contact for further information): Mr James Anderson	Telephone No.: 041-204 2566, ext. 5637 Telex: 777633

Aim of the Project:

To develop a completely integrated floating production system, designed for optimal development of small North Sea fields.

Project Description:

The project set out to develop a design of a semi-submersible to be purpose orientated for production to overcome the incompatibilities of converted drill rigs and to investigate the possibility of reducing downtime due to environmental conditions by optimising the column stabilising characteristics in association with innovative technology in riser and mooring systems for semi-permanent location without the need for transit considerations.

The study includes the investigation of subsea templates and manifolds with the associated control and instrumentation systems, clustering of satellite wells and the problems of commingling of production flow. Various configurations for the FPF have been considered, e.g. producer only, driller and producer, producer with simultaneous work-over, producer with consecutive workover all with or without storage and a comparison made of the economic viability of these various alternatives.

State of Project:

The first phase has been completed and nine volumes of the results are ready for submission with summaries.

Results and applications:

The first phase has shown clearly the way for optimising production viability of small fields from a semi-submersible. However, the second phase is attempting to investigate some aspects in more detail to give increased confidence in the complete integrated system.

Title : PRODUCTION IN THE ARCTIC SEAS	Project n° : 04.04/76
Contractor : GERTH Address : 4, av. de Bois Préau 92500 RUEIL-MALMAISON FRANCE Technical director (or person to contact for further information) : Mr Gilbert BLU or Mr Bernard MERCIER	Telephone n° : (1) 749.02.14 ext. 2288 or 2747 Telex : 203 050 F

AIM OF THE PROJECT

The purpose of the project is to study the protection of the production facilities in arctic zones from scouring by icebergs.

PROJECT DESCRIPTION

This scheme is based on the subsea installations that are set down into excavations made in the sea floor, protecting them from scour by the icebergs. They are interconnected by flexible lines which would not endanger the installations if they were accidentally torn away by an iceberg. Production control and its evacuation by tankers is ensured from a dynamically positioned floating cylinder production platform equipped with devices preventing it from being carried along by the pack ice on the surface.

The activities have been directed towards the key technical factors namely :

- the presence of very hard moranic boulders is an obstacle to drilling the wells. A seismic exploration campaign to detect the zones of boulders at present below the sea bed was made on several sites in Sea of Labrador, though without enabling a reliable method of detection to be determined.
- an experimental excavation in a depth of 150 metres of water was made on the Hermine site of Britany. This 1 000 m3 excavation was made in 20 hours from the "Duplus" ship under average conditions of weather. It was completed by a shore test in a soil containing boulders, using the same excavator bin.
- the production installations are based on using "insert tree" production equipment buried to a depth of over 15 metres beneath the sea floor. The remote-handling control and flowline connection devices are mounted above the well inside a large diameter insert, itself located over 5 metres below the surface of the sea bed. An engineering file was established.
- lastly, the dynamically positioned cylindrical production platform, know as the "Dypospar" was studied. This study, which has received the approval of the certification authorities, also includes study of the production riser.

STATE OF THE PROJECT

The project began on 1st of March 1974 and was ended on 30th June 1979.

RESULTS AND APPLICATIONS

All this work has enabled the feasibility of the seasonal production scheme adopted to be demonstrated, though without going beyond this, since no actual application was presently available for it.

Title: Separation processes	Project No.: 04.08/76
Contractor: The British Petroleum Co. PLC Address: Britannic House, Moor Lane, UK - London EC2Y 9BU Technical director (or person to contact for further information): D.L. Knights	Telephone No.: (01) 920 80000 Telex: 888811

Aim of the Project:

Work on this contract was aimed at the reduction in size and weight of gas/oil and water/oil separators thus reducing the cost of offshore platforms.

Project Description:

The gas/oil separation development concentrated on the use of a cyclone. A system using two cyclones in series was developed and tested, first in the laboratory, then at pilot scale in Kuwait at a throught-put of 10 000 barrels/day. The unit gave specification performance with the addition of anti-foam agent.

A system based on the use of cartridge coalescers protected by a backflushable prefilter was developed for the dewatering of crude oil and subsequent effluent water deoiling. The system was tested in the East Midland oilfield at 400 barrels/day and attained the targets of less than 0.5 % volume water in oil and less than 15 ppm oil in water.

State of Project:

This project was completed on 28 February 1978 and was reported in June 1978.

Results and applications:

Sufficient information was obtained to enable the design of prototype separators to proceed with confidence.

SECONDARY AND ENHANCED RECOVERY

Title: Research on oil recovery from heavy oil deposits under the Adriatic Sea	Project No.:11/75
Contractor: Agip SpA + Deutsche Shell Address: P.O. Box 12069 I - 20230 Milano Technical director (or person to contact for further information): Prof. G.L. Chierici	Telephone No.: (2) 520 40 86 Telex: 310246 ENI I

Aim of the Project:

The purpose of the project is to further explore the possibilities of developing the heavy oil in fractured limestones of Upper Cretaceous-base Tertiary ages of the Adriatic Sea, offshore Pescara.

Project Description:

Drilling, coring and testing of a well in the Emilio structure. Analysis of cores and fluids obtained from said wells. Laboratory studies to initiate the development of a suitable recovery method.

State of Project:

Completed.

Results and applications:

Well Emilio 4 was spudded in on June 3, 1976 and reached the final depth of 3 400 m. on Sept. 20, 1976. It was continuously cored from 2 890 to 3 240 m: the cored section was found practically tight, with a few fractures only showing stains of bitumen. Seven drill-stem tests and one production test were performed with disappointing results. A corrosion cap was installed, and the well abandoned.

Title: Improved crude oil production and treatment	Project No.: 05.01/76
Contractor: The British Petroleum Co. PLC Address: Britannic House, Moor Lane, UK - London EC2Y 9BU Technical director (or person to contact for further inofrmation): D.L. Knights	Telephone No.: (01) 920 80000 Telex: 888811

Aim of the Project:

The overall objective of the project was to increase the amount of usable crude oil produced during primary and subsequent recovery phases in North Sea reservoirs.

Project Description:

To test novel and commercially available displacement fluids in laboratory and field trials for potential recovery of in situ crude oil. Further to test under laboratory and field conditions whether such recovered crude oils (often in the form of water-in-oil emulsions) can be satisfactorily dehydrated and desalted on an offshore production platform.

State of Project:

The project was completed in 1978. Laboratory equipment and pilot rigs have been constructed to evaluate the displacement fluid efficiencies and the dehydration of (simulated) recovered oil. Evaluations have been completed and some field trials carried out to confirm laboratory predictions.

Results and applications:

A range of in-reservoir oil displacement fluids was investigated. In particular a microemulsion fluid was developed to raise the injectivity of water injection wells by displacing flow impeding residual crude oil out of the critical flow zone arund the well perforations. This development was continued through rigorous laboratory simulation tests and large-scale plant blending to a successful full-scale test with a Forties field injection well. Other development of surfactants and temperature sensitive polymer solutions was terminated at the laboratory test phase.

Work on identifying the optimum conditions for achieving the separation of co-produced water from crude oil was carried out within offshore platform restrictions. A unique test rig was built and successfully used to identify optimum conditions and chemical additives for North Sea crude oils. Work culminated with a field test (onshore) at the Forties oil treatment plant, substantiating test rig results. The function of the test rig and associated laboratory test procedures has been explained to many European manufacturers of additives. The problems associated with the production of surfactant and polymer contaminated crude oils that might arise from tertiary recovery operations have been investigated.

An associated programme of more fundamental research supported the project.

Title : PILOT INJECTION OF MICROEMULSION AND POLYMERS IN THE CHATEAURENARD RESERVOIR	Project n° : 05.02/76
Contractor : GERTH Address : 4, av. de Bois Préau 92500 RUEIL-MALMAISON FRANCE Technical director (or person to contact for further information) : Mr Gilbert BLU or Mr Bernard MERCIER	Telephone n° : (1) 749.02.14 ext. 2288 or 2747 Telex : 203 050 F

AIM OF THE PROJECT

The aim of the project was to evaluate a method of oil recovery based on the injection of a plug of microemulsion soluble in the oil and the water of the field, subsequently displaced by a concentrated solution of hydrosoluble polymers.

PROJECT DESCRIPTION

The project consisted of the following tasks :

- laboratory studies to derive a formula for the solutions, test their efficiency and select the most appropriate one,
- additional reservoir studies to select the position of the pilot project so as to carry out solution injectivity tests,
- execution of an injection pilot into Chuelles reservoir (Chateaurenard permit), using a configuration of inverse "five spots" with sides measuring 100 metres and an intermediate observation well between the injection wells and the production well. The reservoir lies at a depth of 600 metres and consists of a level of non-consolidated sandy clay 3 metres thick, containing a relatively viscous oil (40 centipoises).

The slug of microemulsion (964 m3) was injected from 7th February to 14th March 1978, and injection of concentrated polymers was started on March 20th, continuing to the end of the year.

On 31st October 1979 6 958 m3 of polymers have been injected out of a total of 7 922 m3 representing approximately 80 % of the pore volume of the layer. Recovery (primary and tertiary) represented about 55 %. This figure must be considered with reservation, since it is difficult to distinguish the product originating from within the pilot panel.

STATE OF THE PROJECT

The work started in July 1975 and ended on 31st October 1979.

RESULTS AND APPLICATIONS

This methodology pilot project represents a technical success, though the quality of the oil produced compared to the cost of the products injected shows that the method is not yet acceptable from the economic standpoint. New work using new products and a new, wider, well spacing habe been employed within the framework of contract 05.21/80.

Title: Development of heavy oil processes	Project No.: 05.03/76
Contractor: Wintershall AG Address: Friedrich-Ebert Strasse 160 D - 3500 Kassel Technical director (or person to contact for further information): V. Gojo	Telephone No.: 0561 301 81 94 Telex: 99632

Aim of the Project:

The development of a process for recovery of heavy oil in West Germany. The Nordhorn field was discovered in 1942, but oil is not recoverable by conventional means.

Project Description:

In 1975, 5 wells were drilled, which should give a clear picture of the geology and dimensions of the reservoir. Most important, they will give rock and oil samples from all wells as they have all been completely cored. Very little oil has been recovered form the cores; all the wells were perforated and 300 litres of oil were recovered from 2 wells. This oil had accumulated in the bore hole over a period of some weeks. The laboratory tests carried out were:

- evaluate the rock physics data
- evaluate oil properties
- flooding tests with hot water and vapour
- tests of in situ combustion
- solvent tests with organic solvent
- research into uses of the oil.

Because of the high viscosity, 1 000 000 centipoises at reservoir conditions, only thermal methods and the use of solvents seem appropriate. At well Nordhorn 1005, two steam injection tests (November/December 1977) had to be abandoned after four days because of too small absorbtion by the reservoir. The second injection test (July/August 1978) was applied using higher pressures, and the well absorbed within 14 days 1 000 t of vapour and the production test obtained 56 t oil and 290 m3 water.

Resulsts and applications:

No further field tests at Nordhorn are envisaged until a better understanding of the uses of the oil is obtained. The following possibilities will be looked at:

1) use as fuel in power stations
2) raw material in the manufacture of bitumen
3) raw material in coking fuels.

Tests carried out so far have shown that steam flooding and steam stimulation can be possible recovery methods. However, it is not yet clear whether the oil can be economically produced.

Title: Research on improved hydrocarbon recovery from chalk deposits	Project No.: 05.04/76
Contractor: Shell Internationale Petroleum Maatschappij BV Address: Carel van Bylandtlaan 30 NL - 2501 AN Den Haag Technical director (or person to contact for further information): T. L. van Waart	Telephone No.: (070) 779 111 Telex: 31005

Aim of the Project:

To develop and field test methods for improvement of hydrocarbon recovery from tight chalk reservoirs, resulting in a reduction of the field development costs per barrel of produced oil.

Project Description:

The three main areas of investigation are:
- development of a stimulation technique which will give sustained improvement in productivity of soft chalk rocks
- development of a drainhole drilling technique which will increase the effective inflow area from the chalk into the wellbore
- development of improved productivity evaluation techniques, aimed at better identification of productive zones and of potentially productive zones requiring stimulation in a hydrocarbon bearing chalk deposit.

State of Project:

The project was terminated on 31 December 1980.

Results and applications:

A new stimulation technique has been developed for chalk formations. This technique, consisting basically of a combined acid/propped fracturing treatment, gives a higher sustained productivity improvement in soft chalk reservoirs than a conventional acid or propped fracturing treatment.

Several wells in North Sea chalk reservoirs have been stimulated with the propped acid fracturing technique; productivity improvements in general close to prediction and production levels showed no strong decline.

A comparative study of alternative drainhole configurations showed that drilling ultra-high-angle wells in a chalk reservoir and selectively stimulating those wells with the new technique could substantially reduce the drilling and completion costs per barrel of produced oil.

Methods have been developed to predict the chalk matrix productivity and the orientation of natural and hydraulically induced fractures in a chalk reservoir. A fracture-detection logging tool, the Circumferential Acoustic Device (CAD), has been constructed. Laboratory investigations showed a promising performance. During field testing the tool operated with encouraging response to formation characteristics and repeatability.

Title: Exploitation of bitumenous shales	Project No.: 05.05/76
Contractor: GERTH Address: 4, avenue de Bois Préau F - 92500 Rueil-Malmaison Technical director (or person to contact for further information): Mr Gilbert Blu or Mr Bernard Mercier	Telephone No.: (1) 749 02 14 ext. 2288 or 2747 Telex: 203050

Aim of the Project:

Tar shales are sedimentary rocks, highly enriched (>5%) in organic matter (kerogene) and hence more or less combustible. Accordingly, the aim of the project was to determine their ability of produce hydrocarbons by distillation.

Project Description:

Two technologies were studied in laboratory conditions:

Technology of pyrogenation

Tests were carried out under various atmospheres and with various heating programmes to determine the optimal pyrogenization conditions for the shale of the Paris Basin. For a series of conditions retained as the most promising, pyrogenization of significant quantities were carried out for valorization studies. Pyrolyses tests on a very tall pilot installation were effected by a contractor leading to determination of the necessary elements for a technical and economic evaluation. Owing to its characteristics and its situation in Europe, the Lurgi-Ruhrgas process was retained rather than the others. Complementary tests were carried out in the laboratory to test a variant of the Lurgi process.

Technology of in situ exploitation

The process already tested by Occidental Petroleum consists of previous mining of 15 to 20% of the layer. A charge is exploded, followed by an underground combustion in the caved-in zone of the chamber created by the explosion.

The programme implemented a study of combustion of shales in grain form and in block form and a study in a pilot area of combustion and pyrolysis of heavily crushed shale.

State of Project:

The project started on 1 April 1975 and ended on 31 December 1977.

Results and applications:

Highly sulphurated and nitrated shale oil is strongly unsaturated, the yields higher in middle distillates and lower in the heavier distillates. Concerning the Lurgi tests, it would appear that the yields of oil have been less than anticipated. In situ methods are definitely less advanced than ex situ methods. The combustion tests on grains (= 5mm) and blocks (= 250mm) were exploratory and it was obviously not possible to determine the optimal aeraulic and thermal conditions for processing Toarcia shale.

Title: Development of brine soluble polymers and associated chemicals for the enhanced recovery of petroleum	Project No.: 05.06/77
Contractors: The British Petroleum Co. PLC Address: Britannic House Moor Lane UK - London EC2Y 9BU Technical director (or person to contact for further information): D.L. Knights	Telephone No.: (01) 920 8000 Telex: 888811

Aim of the Project:

To develop a polymer system for enhancing oil displacement efficiency which will be suitable for the saline, high temperature conditions encountered in European reservoirs.

Project Description:

By decreasing the mobility of water in a reservoir, a small proportion of dissolved polymer can reduce the extent to which oil is by-passed because of the high viscosity of the oil or because of the existence of relatively high permeability channels in the reservoir. Previous laboratory research and field tests on polymer flooding had been restricted to low salinity, low temperature conditions and so the main task was to ensure that polymers could be made to operate under European reservoir conditions for long periods. The possibility of augmenting the polymer flooding process with the aid of chemical additives such as surfactants was also investigated.

State of Project:

The work was completed according to schedule and published.

Results and applications:

Conventional poly(acrylamide) mobility control agents were shown to have inadequate stability and poor flow behaviour in porous media. Poly(vinyl pyrrolidone) was the only synthetic polymer with sufficient stability, but its viscosifying properties were unacceptable. Purified scleroglucan biopolymer gave excellent viscosifying properties in sea water at 90°C without significant deterioration for more than a year. The development was not taken beyond the laboratory stage because of difficulty in obtaining sufficient purified polymer in a form suitable for injection and because there was no suitable opportunity for a field trial.

Title : PILOT FOR HEAVY OIL BY STEAM INJECTION UPPER LACQ OIL FIELD	Project n° : 05.07/77
Contractor : GERTH Address : 4, av. de Bois Préau 92500 RUEIL-MALMAISON FRANCE Technical director (or person to contact for further information) : Mr Gilbert BLU or Mr Bernard MERCIER	Telephone n° : (1) 749.02.14 ext. 2288 or 2747 Telex : 203 050 F

AIM OF THE PROJECT

Steam injection is the most technically proved EOR method, with numerous developments at field level in heavy oil deposits mainly in USA, Venezuela and Canada.
However, all these operations are implemented in sandstone reservoirs and there was no experience of steam injection in carbonated reservoirs which are, in addition, often fractured ones.
The Lacq Superieur project was aimed at testing steam injection methods in this kind of heterogeneous carbonated reservoir.

PROJECT DESCRIPTION

Upper Lacq oil field, discovered in 1949, exhibits very heterogeneous facies : either porous calcareous rocks with a low permeability, or dolomites with a low porosity but a high permeability through the fissure network.
In the fractured zones, after an initial high anhydrous production rate, the wells showed a fast decline in oil production when water from the underlying aquifer invaded them. After a laboratory research program, steam injection was selected as the most favorable for process.
A reservoir study and reference field tests were performed to choose a pilot site and to define operating conditions. A steam injection well was drilled, surface equipment was installed, and a steam drive pilot protect started in October 1977.
A stable steam injection rate of 175T/day has been maintained and only three months after injection began, the oil production rate of some wells increased significantly and the maximum oil production was observed in mid 1979.

STATE OF THE PROJECT

The steam injection was maintained till April 1982 after the end of project (30th June 1980). Incremental oil production was 35 600 T and 251 000 T of steam had been injected (OSR = 0.15 V/V).

RESULTS AND APPLICATIONS

The results showed a good efficiency of heat transfer inside the reservoir in spite of an irregular sweeping of the pilot area and the carbonate decomposition by steam.
A numerical reservoir description, with a "fissure-matrix" configuration, enabled simulation results to be improved.
This steamdrive pilot is considered as a technical success and the steam is now being injected in two new patterns.

Title: Use of EOR processes in the Cortemaggiore field, Italy	Project No.: 05.08/77
Contractor: Agip SpA Address: P.O. Box 12069 I - 20120 Milano Technical director (or person to contact for further information): Prof. G.L. Chierici	Telephone No.: (2) 520 4086 Telex: 310246 ENI I

Aim of the Project:

The purpose of the project is to evaluate different EOR processes potentially applicable to the depleted oil field of Cortemaggiore, a conventional light oil reservoir exploited since 1949.

Project Description:

A new reservoir study, carried out using a synergetic reservoir geology/reservoir engineering approach, laboratory studies on reservoir oil additioned with CO2, N2, natural gas, displacement tests on cores at reservoir conditions using miscellar/polymer cushions and numerical model simulations were carried.

State of Project:

Completed.

Results and applications:

Carbon dioxide miscible displacement and vaporisation of the residual oil into cycled natural gas were the only EOR processes which resulted as technically feasible. Miscellar/polymer processes can be applied in principle, but the surfactant able to cope with the high-salinity reservoir brine is very expensive.

Workovers in some wells, which had been abandoned, and drilling of one infill well resulted in a sizeable production of oil, which is still continuing at present.

Title: Drainage system Rospo Mare (offshore Italy - Adriatic Sea)	Project No.: 05.09/78
Contractor: Elf Italiana SpA Address: Largo Lorenzo Mossa Via Aurelia 619 I - 00165 Roma Technical director (or person to contact for further information): Jean Paul Demesy	Telephone No.: (6) 63890 1 or (6) 63890 323 Telex: 614272

Aim of the Project:

Search for drainage methods for heavy, viscous oil deposits involving risk of water inflow.

Project Description:

1st Phase: Research of a specific model after a complete geological study of reservoir
2nd Phase: Model study of horizontal well efficiency
3rd Phase: Study of CO_2 gas injection
4th Phase: Study of steam flood efficiency
5th Phase: Study of miscellaneous methods to improve oil recovery
6th Phase: Study of miscellaneous methods to delay water arrivals
7th Phase: Concept of a pilot project.

State of Project:

Project realised from 1979 to 1983.

Results and applications:

The first two phases have mainly been developed:

- The geological synthesis resulted in a better understanding of the architectural scheme of a karstic type reservoir and the interference test between wells enabled rock properties to be specified.
- The model study of a horizontal well demonstrated the greater efficiency of such a well in comparison with a vertical well.
- A quick study of CO_2 or steam injection showed that these methods were not promising, compared with the good results obtained from the horizontal well.
- The other methods could not be studied because of lack of sufficient precise data.

Title: New techniques to be applied in Piropo heavy oil field to achieve economical exploitation	Project No.: 05.10/78
Contractor: Agip SpA Address: P.O. Box 12069 I - 20120 Milano Technical director (or person to contact for further information): Prof. G.L. Chierici	Telephone No.: (2) 520 40 86 Telex: 310246 ENI I

Aim of the Project:

The purpose of the project of the project is to evaluate different well stimulation and EOR techniques which showed a potential capability of bringing the oil production rate of Piropo wells to a commercial value.

Project Description:

Sismic data were reprocessed by the Sismolog method to locate the best (most fractured) area of the reservoir, where a well was drilled. A MHF job was carried out in the most promising interval of the well. Core and PVT studies were carried out to evaluate productivity improvement by huff-n-puff with CO_2.

State of Project:

Completed.

Results and applications:

Due to the very low permeability of the reservoir rock, no commercial production was obtained from the well, in spite of the MHF carried out. The stabilised rate obtained was 70 BOPD only. Huff-n-puff with carbon dioxide is promising, in principle, but the high cost for carrying out such EOR process in an offshore oil accumulation rules out any possibility of economic success.

Title : IMPROVING RECOVERY FROM VERY HEAVY OIL RESERVOIRS	Project n° : 05.11/78
Contractor : GERTH Address : 4, av. de Bois Préau 92500 RUEIL-MALMAISON FRANCE Technical director (or person to contact for further information) : Mr Gilbert BLU or Mr Bernard MERCIER	Telephone n° : (1) 749.02.14 ext. 2288 or 2747 Telex : 203 050 F

AIM OF THE PROJECT

The aim of the project was to evaluate the feasibility of combining solvent injection with thermal recovery methods for producing extra-heavy oils.

PROJECT DESCRIPTION

Phase 1 of the project concerned studies outside the porous media :

- . to determine the amount of asphaltenes precipitated when solvents are added to extra-heavy crudes.
- . to measure the viscosity of solvent-crude oil mixtures.
- . to recommend solvents technically and economically acceptable, for use in oil reservoirs.

Phase 2 of the project involved displacement tests in special one dimensional cells, to evaluate the efficiency of :

- . combined injection of solvents and steam,
- . in situ combustion combined with solvent injection.

STATE OF THE PROJECT

The project began on 1st February 1978 and was terminated on 30th June 1981.

RESULTS AND APPLICATIONS

- . a correlation has been obtained to calculate the viscosity of crude oil/diluent mixtures with relation to dilution and temperature.
- . data has been obtained on the effect of the composition of the diluent on asphaltene precipitation.
- . the injection of a solvent slug in combination with steam can lead to decreased residual oil saturation ; on the other hand, the addition of solvent led to decreased air requirements only in a limited number of cases.
- . the main benefit of solvent injection is to improve the subsequent injectivity to a displacing agent.

Title: Development of an electrical system for preventing coning of water into producing oil wells	Project No.: 05.12/78
Contractor: SYMINEX Address: 2, Boulevard de l'Océan F - 13275 Marseille Cédex 9 Technical director (or person to contact for further information): A.J. Kermabon	Telephone No.: (91) 73 90 03 Telex: 400563

Aim of the Project:

From 1976, SYMINEX started experimenting on the effects of DC current, when applied to oil water saturated rocks (based on previous work done by TIKHOMOLOVA: Un. of Leningrad - CHILLINGAR: Un. of Los Angeles - HEADLEY and PIERCE: US Bureau of Mine..., which basically showed that fluids can be moved in a porous medium with the help of DC current). The aim of this project was initially to study a possible application of the mechanism for water coning control on producing wells. However, a new approach of the mechanism observed at laboratory level motivated SYMINEX to reorient the project towards a different application which was EOR.

Project Description:

The main aspects of work done during this contract were:

- laboratory experiments to help the understanding of physical mechanism; these experiments were conducted at the Ecole Centrale de Paris
- technical studies to evaluate the feasibility of an extension of the process to the oil reservoir (EOR).

State of Project:

The reorientation of this project will lead SYMINEX to the end of this contract. However, a new contract relative to EOR is in progress.

Results and applications:

The main results were:

- low amount of current needed to electrically stimulate the oil recovery out of a rock sample saturated with oil and water
- injecting electricity (about 1 000 amps DC) down to 1 000 metres appears feasible.

This process seems to be particularly suited for fractured reservoirs, where the usual techniques based on hydraulic flooding and chemical injection have only mediocre results.

Title: Pilot plant to enhance heavy oil recovery. Ponte Dirillo field, Italy	Project No.: 05.14/79
Contractor: Agip SpA Address: P.O. Box 12069 I - 20120 Milano Technical director (or person to contact for further information): Prof. G.L. Chierici	Telephone No.: (2) 520 40 86 Telex: 310246 ENI I

Aim of the Project:

Design, build and put into operation a field pilot for a EOR process by non-miscible CO_2-rich gas injection into the very heavy (11 to 17° API) oil field of Ponte Dirillo, Italy.

Project Description:

The project is based on the results of a previous study, carried out on the heavy oil field of Gela (Contract 16/75) of similar characteristics. A campaign of well surveys, laboratory tests and a numerical model simulation were carried out. The pilot plant was designed, built and put into operation.

State of Project:

Almost completed.

Results and applications:

According to numerical model results both peripheral and crestal injection of high-pressure, CO_2-rich natural gas can sizeably increase oil recovery.

Two wells have been completed for gas injection, which started in December 1983 in the peripheral well No. 9. A 50% increase in oil production rate of the field (from 110 to 170 cu m/d) resulted from this gas injection. The injection of gas is now being switched to the crestal well No. 5, which will be operated in the huff-n-puff mode.

Title: Seismic and geoelectric methods for FRAC-location from the surface	Project No.: 05.15/79
Contractor: Prakla-Seismos GmbH Address: Buchholzer Str. 100, D - 3000 Hannover 51 Technical director (or person to contact for further information): Dr E Wierczeyko	Telephone No.: (0511) 642-0 Telex: 922847 922419 923250

Aim of the Project:

Determination of the propagation direction and propagation distance of hydraulic fracs produced in low permeable oil/gas deposits using seismic and geoelectric methods.

Project Description:

During hydraulic fracturing of oil/gas horizons in the subsurface seismic-acoustic signals are produced indicating the propagation direction and propagation distance of the frac. Recording the change of the electric field in the frac region by active and passive geoelectric measurements at the surface can give information about the propagation direction of the frac.

State of Project:

Project can be completed by the end of 1984.

Results and applications:

The propagation direction of hydraulically produced fracs can be determined by active and passive geoelectric measurements.

The propagation distance can be determined by recording seismic-acoustic signals generated during fracing, as long as three component acceleration receiver sondes in auxiliary boreholes near the frac can be applied. After completion of a suitable receiver sonde it will be attempted to record, in the frac borehole itself below the fracpoint, the seismic-acoustic signals generated during and after fracturing and consequently to determine at least the propagation distance, i.e. the frac length.

Title: Enhanced oil recovery from the Egmanton oil field by carbon dioxide miscible flooding	Project No.: 05.15/80
Contractor: The British Petroleum Co. PLC Address: Britannic House Moor Lane UK - London EC2Y 9BU Technical director (or person to contact for further information): D.L. Knights	Telephone No.: (01) 920 8000 Telex: 888811

Aim of the Project:

The project was established to evaluate the feasibility of a carbon dioxide miscible flood within the West European area and to conduct a pilot flood as a tertiary process in the watered-out Egmanton reservoir.

Project Description:

The project comprised laboratory and computer modelling studies (Phase I) and preparation for, and operation of, a five well pilot test (Phase II) together with further studies. BP entered into agreement with Compagnie Française des Pétroles, Société Nationale Elf-Aquitaine (Production) and Institut Français du Pétrole, whereby these French partners performed part of the studies and met 50% of the total expenditure on the studies (less the appropriate EEC grant).

State of Project:

The project is terminated after completion of studies, engineering work, drilling, well workover and production testing and a sequence of water and CO_2 injection tests.

Results and applications:

All data required for simulation and engineering design was generated in the laboratory studies and an understanding of the CO_2 miscible flood mechanism was built up. After initial difficulties it eventually proved possible to model the miscible flood process in the Egmanton pilot area using two approaches. Within Phase II, a programme of drilling, workover and individual well tests, and two interference tests, provided core material and data for the field model. The site and facilities required for the pilot project were designed, built and commissioned. They were then used to test production wells and to inject CO_2 and water on a test basis.

However, low CO_2 and water injectivity confirmed the decision to terminate the project prematurely on the basis of the extended time scale - as indicated by the model studies - and the escalating cost to complete the pilot test. A reduced scale pilot test scheme was not found viable. Project activity has formed the basis of four publications.

Title: Downhole steam generator for enhanced oil recovery	Project No.: 05.16/80
Contractor: The British Petroleum Co. PLC Address: Britannic House Moor Lane UK - London EC2Y 9BU Technical director (or person to contact for further information): D.L. Knights	Telephone No.: (01) 920 8000 Telex: 888811

Aim of the Project:

To develop a downhole steam generator for the thermal recovery of crude oil to the stage of initial preparations for a field trial. The generator incorporates a pulsed burner, patented by BP, which should overcome the materials problems arising from the sustained high temperatures which occur in continuous burners under these conditions.

Project Description:

Steam for oil recovery is conventionally injected from generators mounted at the surface; a downhole steam generator would eliminate heat losses from the wellbore and may offer other advantages, such as greater flexibility in the siting of surface equipment, reduced feed water treatment and less atmospheric pollution. The BP generator also injects the products of combustion, carbon dioxide and nitrogen, with the steam. Thus flue heat losses are eliminated, and there is evidence that these additional gases can increase the oil recovery rate.

State of Project:

A 5 MW (1 000 barrel/day of steam, cold water equivalent) prototype generator has been designed and parts of the unit manufactured. Detailed planning has been carried out for a possible field test in the UK. This contract has been completed and further development will be influenced by parallel technical and economic studies.

Results and applications:

The design of the prototype generator incorporates the results of an extensive development programme, chiefly on the novel pulsed high pressure combustor, the ignition system and the engineering of the burner tube. A rig has been built to test reduced scale combustors at up to 70 bar.

Potential applications include: fields where co-injection of exhaust gases could enhance recovery rates; offshore locations where equipment siting and water treatment could be important factors; permafrost zones where melting around the injection well has to be prevented. The development of insulated tubing by others has provided an alternative and possibly more attractive method of injecting steam into deep reservoirs. However, a key factor in all cases is the extent of the benefit form the co-injection of exhaust gases. Substantial benefit would favour this type of generator.

Title : MARIENBRONN INJECTION PILOT PROJECT	Project n° : 05.19/80
Contractor : GERTH Address : 4, av. de Bois Préau 92500 RUEIL-MALMAISON FRANCE Technical director (or person to contact for further information) : Mr Gilbert BLU or Mr Bernard MERCIER	Telephone n° : (1) 749.02.14 ext. 2288 or 2747 Telex : 203 050 F

AIM OT THE PROJECT

The MARIENBRONN injection pilot project is intended to master the technology and methodology of exploitation of heavy oils by steam injection.

PROJECT DESCRIPTION

The field of application is MARIENBRONN reservoir, which contains an oil with a density of 0.99 and an in situ viscosity of 227 poises. This reservoir consists of sands and conglomerates and lies about 250 metres below the surface. It has a monocline type structure.

19 wells have been drilled in a "five spot" configuration on the MARIENBRONN pilot project.

The production equipment for the project comprises feed water processing units, a steam generator, separation and storage tanks and a plant for treating oily waters. After purification, the production waters are discharged into a well in the geological stratum.

STATE OF THE PROJECT

After a number of difficulties encountered in introducing the "forced circulation" type steam generator into service had been solved, the first stimulation cycles by huff and puff of the various wells took place in 1984. The initial results of the injections revealed the highly heterogeneous nature of the reservoir and establishment of communications between certain wells shows that, in these sectors, the stimulation phase must be replaced directly by the steam drive phase.

When introducing the production equipment into service, serious difficulties were encountered with the oily water treatment system (coalescer and active carbon filters).

Despite the modifications made by raising the operating temperature and controlling the flows, the system is not operating satisfactorily. De-oiling of the production waters by coalescence of drops of heavy oil is not operational.

RESULTS AND APPLICATIONS

Under these conditions, the discharge of the "de-oiled waters" into the geological stratum is problematical and in the medium term, it is expected that the discharge well will be blocked.

Title: Heavy oil pilot project NORDHORN	Project No.: 05.20/80
Contractor: Wintershall AG Address: Friedrich-Ebert Strasse 160 D - 3500 Kassel Technical director (or person to contact for further information): Günther Proyer	Telephone No.: (0561) 301-8194 Telex: 99632

Aim of the Project:

Evaluation of the feasibility of the recovery of Nordhorn tar oil by steam injection through hydraulically induced fractures.

Project Description:

The pilot project consists of a 5-acre inverted 5-spot pattern with 5 wells (thereof 4 producers and 1 injector). The crude and water treating system has a capacity of 360 tonnes per day of wet oil production.

State of Project:

Planning work (second project phase), including laboratory tests, simulation study and conceptual planning, is finished.

Results and applications:

With steam injection (2 000 b/d) in the injector of a 5-acre 5-spot pattern about 215 000 STB (34 000 m^3) of tar oil can be produced during a period of about 2 years - this represents 23% of the OOIP.

Saleable products can be obtained from the heavy oil of Nordhorn by means of conversion processes.

In view of the low net back price for Nordhorn tar oil (utilising delayed coking) and the corresponding low proceeds, high investments and operating costs, the "Nordhorn Heavy Oil Project" is economically not attractive to date.

Title : INDUSTRIAL PILOT PROJET FOR INJECTION MICROEMULSION AND POLYMERS INTO NEOCOMIAN RESERVOIRS (CHATEAURENARD)	Project n° : 05.21/80
Contractor : GERTH Address : 4, av. de Bois Préau 92500 RUEIL-MALMAISON FRANCE Technical director (or person to contact for further information) : Mr Gilbert BLU or Mr Bernard MERCIER	Telephone n° : (1) 749.02.14 ext. 2288 or 2747 Telex : 203 050 F

AIM OF THE PROJECT

The feasibility state of this industrial pilot project comprised a combination of tasks for defining the specifications and specificity of the project in the light of the results obtained on the methodology pilot project executed under contract 05.02/76, namely :

- the search for an economic return with the elaboration of less expensive microemulsion solutions and the optimization of the volume to be injected
- elaboration of an industrial scale project
- the dimensioning of the panels and the choice of the techniques for preparing and injecting solutions.

PROJECT DESCRIPTION

- laboratory studies to define the nature and volume of the products to be injected
- additional reservoir studies so as to position the pilot project panel and to characterize the reservoir by means of interference tests
- drilling and completion of 5 new wells
- design of units for producing and injecting the solutions preparing the ground for industrial exploitation stage.

STATE OF THE PROJECT

The work started in October 1979 and ended on 30th June 1981.

RESULTS AND APPLICATIONS

The interference tests confirmed the soundness of the site chosen by proving the continuity of the reservoir with no high contrast anisotropy of the permeabilities.

Phase 1 of the programme of work can only be considered as a partial success. On completion of the project, the laboratories proposed a formula, the sulfonate content of which has for obvious economic reasons been halved compared to that of the methodology pilot project covered by contract TH 05.02/76. This formula has proved its efficiency in a non-clay medium, but is not yet perfectly suited to the conditions of use at Chateaurenard reservoir. Studies of a new microemulsion were hence continued outside the scope of the contract and the injection of this new microemulsion is covered by contract TH 05.28/81.

Title: Upgrading of heavy crude and bitumen	Project No.: 05.22/81
Contractor: Veba OEL AG Address: P.O. Box 45 Alexander von Humboldt Str. D - 4650 Gelsenkirchen-Hassel Technical director (or person to contact for further information): Dr. Holighaus	Telephone No.: 0209/366 79 68 Telex: 824881-0 vo d

Aim of the Project:

Development of an upgrading process for heavy crude and bitumen into storable and transportable synthetic crudes.

Project Description:

The key element in the process development for heavy crude and bitumen upgrading is the construction and operation of a pilot plant with a nominal capacity of 1 t/h oil input. This project covers only the work until the plant was mechanically completed and commissioned.

State of Project:

Mechanical completion was reached in April 1983, commissioning and start up took place until end of May 1983.

Results and applications:

The overall investment of the pilot plant including engineering and training of the operators amounts to 47.8 Million DM. The start-up procedure already indicated that the plant has been properly engineered. The results to be generated will be used for process design and optimisation of a commercial plant. The operation of the pilot plant is performed under a separate EC contract.

Title: Method of oil recovery by carbon dioxide injection in the Coulommes-Vaucourtois field	Project No.: 05.23/81
Contractor: Petrorep Address: 42, ave Raymond Poincaré, F - 75116 Paris Technical director (or person to contact for further information): M. E. Couve de Murville	Telephone No.: 16 (1) 505 14 00 Telex: 611 036 F

Aim of the Project:

(1) Check that the upper reservoir, of very low permeability, can be produced by carbon dioxide injection and (2) Find out if the production of the nearly depleted lower reservoir can be extended economically by injecting carbon dioxide and nitrogen.

Project Description:

The project is divided into three phases:

Phase I: Theoretical studies including a geological updating of the structure with emphasis on fracture system, a laboratory study on miscibility and fluids behaviour, a reservoir analysis based on production data and simulation;

Phase II: Injectivity and production test in the upper reservoir of the same well;

Phase III: Test the displacement efficiency and profitability with a four-spot pattern in the lower reservoir.

State of Project:

Phases I and II have been completed in 1981 and in 1982; Phase III is progressing. 2600 tons of carbon dioxide and 900 tons of nitrogen were injected in 1984. The production through nearby wells and the overall appraisal of the project will extend at least to the end of 1985.

Results and applications:

Theoretical studies, lab measures and simulations (Phase I) showed that, at bottom hole conditions, the miscibility oil/carbon dioxide is effective and that additional oil recovery by displacement could be satisfactory.

The injection-production test through the same well in the low permeability upper reservoir (Phase II) has proved disappointing and the well shut down a few years ago. At time of writing it is too early to draw any conclusions on the displacement of oil in the lower reservoir (Phase III) as no sign of breakthrough nor oil increase have been noticed in the nearby wells.

Title: Production test of Rospo Mare oil field (offshore Italy - Adriatic Sea)	Project No.:05.24/81
Contractor: Elf Italiana SpA Address: Largo Lorenzo Mossa 8 Via Aurelia, I - 00165 Roma Technical director (or person to contact for further information): Mr Jean Paul Demesy	Telephone No.: 39 6 63890 1 or 39 6 63890 323 Telex: 614273 ELFMIN I

Aim of the Project:

To study water breakthrough and water cut evolution at Rospo Mare field, a heavy and viscous oil field with bottom water drive.

Project Description:

1st stage: realisation and laying of a fixed platform "Rospo Mare A" with its evacuation terminal. Drilling of 2 partial penetration wells: a vertical one, a directional one (out of project). Realisation of production equipment.
2nd stage: production test with interference test, and study of water breakthrough time and water cut evolution.

State of Project:

First stage was realised in 1981/beginning of 1982. Production test (Stage 2) started in August 1982 and is still continuing (1st half of 1984). The project itself ended in December 1982.

Results and applications:

- Rospo 4, vertical well, was pumped at 230 m3/d with pressure interference measured on Rospo 5 before pumping this well at 50 m3/d (due to its worse productivity).
- At 230 m3/d, the breakthrough occurred on Rospo 4 after an oil production of 8,700 m3 (in agreement with the value forecast by model): then an initial critical rate of 130 m3/d with reversibility of water coning was evidenced. (Initial critical rate of Rospo 5 was less than 25 m3/D). Even if encouraging, these results do not allow to foresee the development of Rospo oil field with subvertical wells.

Title: Retorting process for Schandelah oil shale	Project No.: 05.26/81
Contractor: Veba OEL AG Address: Alexander von Humboldt Str. D - 4650 Gelsenkirchen-Hassel Technical director (or person to contact for further information): Dr R. Holighaus	Telephone No.: (0209)366 7968 Telex: 8 24 881 - 80 voe

Aim of the Project:

Process development for producing pyrolysis oil and gas from shale utilising residual carbon for heat supply.

Project Description:

Pilot plant with a capacity of 300 kg/h of oil shale. Pyrolysis is carried out first in a cyclone and subsequently in a rotating drum. The heat required for the process is supplied by burning the oil shale residue.

State of Project:

Construction of the pilot plant will be completed at the end of 1984; start of operation is planned for the beginning of 1985.

Results and applications:

Results of the pilot plant shall provide design data for commercial plant.

Title: Pilot plant for increasing the recovery of heavy oil in the Vallecupa field	Project No.: 05.27/81
Contractor: AGIP SpA Address: PO Box 12069 I - 20120 Milano Technical director (or person to contact for further information): Mr Malerba	Telephone No.: (02) 520 4086 Telex: 310246

Aim of the Project:

To develop a pilot plant to enhance the recovery of heavy oil from the Vallecupa field.

Project Description:

This project concerns the injection of steam into the Vallecupa field. A pilot plant will be developed for this purpose. The main phases of the project are:

1. Collection of reservoir and well basic data
2. Study of the reservoir for the selection of the best injection/ production scheme
3. Drafting and construction of the pilot plant (based on a five-spot geometry and with central system injection)
4. Test of the process on the pilot plant.

State of the Project:

- Phase I has been completed.
- Phase II is mostly completed: the detailed seismic interpretation of the structural top of the reservoir has been performed. Petrographical sedimentological analysis of the cores and the stratigraphical study of the whole structure have been concluded. The petrographical evaluation and the study of the fracturing distribution of three continuous corings have been completed.
- Phase III has been completed. The injection and production wells are prepared for the test. Construction of the surface plants has been finished and the plants have been finally tested.
- Phase IV is being started. During some months the injection of steam will be at the highest characteristics (7 tonnes/hour of saturated steam at a pressure of 60 kg/cm^2) to support the reservoir repressurization.

Title : CHATEAURENARD INDUSTRIAL PILOT PROJECT : INJECTION OF MICROEMULSION	Project n° : 05.28/81
Contractor : GERTH Address : 4, av. de Bois Préau 92500 RUEIL-MALMAISON FRANCE Technical director (or person to contact for further information) : Mr Gilbert BLU or Mr Bernard MERCIER	Telephone n° : (1) 749.02.14 ext. 2288 or 2747 Telex : 203 050 F

AIM OF THE PROJECT

The "Chateaurenard industrial pilot project" involves a pilot project for enhanced recovery by injecting microemulsion and polymers. This technique has already been tried and proved on a small scale on the Chateaurenard field methodology pilot project (contract TH 05.02/76). A substantial gain in recovery, estimated at 28 % of the initial accumulation was obtained, but the cost was prohibitive.

The feasibility study of an industrial pilot project performed under contract TH 05.21/80 was based on the use of non-ionic surfactants in the microemulsion formula ; owing to difficulties linked to the slightly clayey sandy lithology of the Chateaurenard reservoir, it was not successfully concluded. Additional studies outside the scope of the contract enabled a microemulsion consisting of 30 % surfactants, 15 % oil, 55 % water to be retained.

To keep this project within its economic objectives, it was decided to inject a very small volume of microemulsion (3.5 % of the pore volume VP).

PROJECT DESCRIPTION

The configuration chosen is a pattern of 4 adjacent inverted 5-spots : on injection well was drilled in the centre of each mesh (average spacing 270 m).

A station was set up in the centre of the panel. Its design features are :
- extensive automation so as to reduce operating costs
- production units by in-line mixing of the solutions to be injected.

STATE OF THE PROJECT

The project was completed on 31st October 1983 with the injection of microemulsion.

RESULTS AND APPLICATIONS

The exploitation schedule and projected plan catered for the injection of :

- 3.4 % of the VP of microemulsion (accomplished in June/July 1983)
- 0.4 % of the VP of concentrated polymer solution (up to October 1984)
- 0.4 % of the VP of diluted polymer solution (up to November 1985)
- 0.4 % of the VP of purified water (up to October 1986).

One will have to wait for the end of the purified water injection phase to decide as to the success of this pilot project.

Title : PILOT PROJECT FOR INJECTION OF MISCIBLE GAS INTO PECORADE RESERVOIR	Project n° :05.29/81
Contractor : GERTH Address : 4, av. de Bois Préau 92500 RUEIL-MALMAISON FRANCE Technical director (or person to contact for further information) : Mr Gilbert BLU or Mr Bernard MERCIER	Telephone n° : (1) 749.02.14 ext. 2288 or 2747 Telex : 203 050 F

AIM OF THE PROJECT

This project concerned the application at PECORADE reservoir of a CO_2 injection pilot project for maintaining the pressure in the reservoir above critical pressure, whilst at the same time improving the oil recovery in a part of the reservoir that was difficult to exploit by conventional methods (compact porous zone, but with low permeability).

PROJECT DESCRIPTION

There were three phases to the programme of work :

Phase 1 - Reservoir study

This comprised the preparation of the following documents :

- a petrophysical model to determine the oil recovery rate
- a compositional model with 4 constituents, to analyse the results of the pilot project
- a method of plotting the CO_2 injected and the appropriate means of analysis in order to distinguish between the CO_2 present in PECORADE oil, and that injected.

Phase 2 - Injection test

Two liquid CO_2 injection tests (20 and 200 tons) were successfully carried out on well PCE 20, with a wellhead pressure of about 500 bars.

Phase 3 - Preparation of the pilot project

The pilot project consists of one injection well PCE 13 and three production wells PCE 04, PCE 22 and PCE 23. The spacing between the wells is about 200 metres.

The surface installations were designed to inject CO_2 at a continous rate of 80 tons per day and a pressure of 500 bars.

The engineering work covered the gaseous and liquid CO_2 production installations located at Lacq plant and the CO_2 de-tanking, storage and injection installations at the PECORADE site.

STATE OF THE PROJECT

The work started on 1st December 1981 and was completed on 31st December 1982.

RESULTS AND APPLICATIONS

The files established during this project were used to build the CO_2 pilot injection installation at PECORADE, outside the EEC support.

Injection of CO_2 into the pilot zone started in March 1983, under contract TH 05.44/82.

Title : PLATFORM FOR PREPROCESSING HEAVY OILS : ENGINEERING	Project n° :05.30/81
Contractor : GERTH - ASVAHL Address : 4, av. de Bois Préau 92500 RUEIL-MALMAISON FRANCE Technical director (or person to contact for further information) : Mr. Jean François LE PAGE	Telephone n° : (1) 749.02.14 ext. 2288 or 2747 Telex : 203 050 F

AIM OF THE PROJECT

This contract concerns the second stage of a project, the intention of which is to perfect methods of preprocessing non-conventional heavy oils that are applicable at the production field, so as to facilitate their conveyance by pipeline.

The first stage of the project concerned the general studies required to define the specifications of the units and to draught the layout drawings.

The second stage of the project concerns the engineering of the installations of the experimental platform, which has a capacity of 20,000 tons per year.

PROJECT DESCRIPTION

The work comprised :
- the engineering and supervision of construction
- follow-up of the construction and acceptance testing of the installations by the operator : ASVAHL (Association pour la valorisation des huiles lourdes).

The field of application of the contract concerned only the thermo-physical method involving the following units : desalination, atmospheric distillation, vacuum distillation, de-asphalting, visbreaking, hydrovisbreaking and the appurtenant installations, namely the utilities, storage facilities and links to the FEYZIN refinery.

STATE OF THE PROJECT

The heavy oil experimental platform was set up at Solaize site.
The engineering work started in March 1981 and the units were fully erected by June 1983.

Title : TAR SHALES : THE TRANQUEVILLE IN SITU COMBUSTION PILOT PROJECT	Project n° :05.31/81
Contractor : GERTH - BRGM - G d F - Address : 4, av. de Bois Préau 92500 RUEIL-MALMAISON FRANCE Technical director (or person to contact for further information) : Mr Gilbert BLU or Mr Bernard MERCIER	Telephone n° : (1) 749.02.14 ext. 2288 or 2747 Telex : 203 050 F

AIM OF THE PROJECT

The aim of the TRANQUEVILLE in situ combustion pilot project was full-scale verification of the feasibility of this method of exploiting the potential energy of tar shales and, if necessary, overcoming the new operational problems arising from this type of exploitation.

PROJECT DESCRIPTION

Two wells were drilled in the layer of shales (depth 200 m - thickness 25 m - distance 60 m). They were in communication by a horizontal crack brought about by hydraulic fracturation from the injection well.

Air was then injected into this crack. Combustion was ignited by electric heating at the bottom of the injection well. The second well enabled the products of combustion/pyrolysis to be collected.

STATE OF THE PROJECT

The work started on 1st July 1980.

A combustion test was performed from 1st December 1983 to 15th January 1984. A core was sampled from the burnt zone. The results are now being analysed.

RESULTS AND APPLICATIONS

Combustion was complete and did not give rise to any production of hydrocarbons. The gases produced always had too high an oxygen content. The method can be applied to in situ heat production, with subsequent recovery of this heat by injection of water.

Title : PILOT PROJECT FOR INJECTION OF STEAM INTO THE "SAINT JEAN DE MARUEJOLS" HEAVY OIL RESERVOIR	Project n° : 05.32/81
Contractor : GERTH Address : 4, av. de Bois Préau 92500 RUEIL-MALMAISON FRANCE Technical director (or person to contact for further information) : Mr Gilbert BLU or Mr Bernard MERCIER	Telephone n° : (1) 749.02.14 ext. 2288 or 2747 Telex : 203 050 F

AIM OF THE PROJECT

The project involves a pilot project for injecting steam into a heavy and viscous oil accumulation in the highly compact and fissurized Saint Jean de Maruejols limestone reservoir.

This pilot project aimed at the following :

- finding out the increase in productivity that can be produced by steam stimulation on the type of highly compact, but fissurized, limestone reservoir
- assessing the possibilities for profitable exploitation of the deposit : an industrial pilot of 5 wells was planned and would have been submitted firstly to a cycle injection, secondly to a draining operation by continuous steam injection through the centre well (steam drive).

PROJECT DESCRIPTION

Between 1947 and 1950 eight wells have been drilled. For this purpose, two new reconnaissance wells were drilled : MAR 101 and MAR 104.

A steam stimulation test on the well MAR 101 took place in three successive huff and puff cycles of unequal duration.

Many operating difficulties were encountered.

Following these three cycles, the deposit pressure dropped rapidly and the well production became pratically nil. The steamed water injected was not recovered.

STATE OF PROJECT

The project was abandoned without carrying out the pilot project (5 spot) as scheduled.

RESULTS AND APPLICATIONS

No significative technical results were obtained from this project.

Title : EMERAUDE PILOT PROJECT FEASIBILITY	Project n° : 05.33/81
Contractor : GERTH Address : 4, av. de Bois Préau 92500 RUEIL-MALMAISON FRANCE Technical director (or person to contact for further information) : Mr Gilbert BLU or Mr Bernard MERCIER	Telephone n° : (1) 749.02.14 ext. 2288 or 2747 Telex : 203 050 F

AIM OF THE PROJECT

This Franco-Italian project consists of a steam injection pilot project on the Emeraude heavy and viscous oil reservoir in the Congo offshore field in a depth of water of 65 metres.

After 7 years of primary exploitation, the reservoir produces less than 3 % of the oil in place and the final recovery with this production method would not exceed 5.2 % of the estimated reserves of 575 million tons.

Application of this thermal enhanced recovery method offshore is a worldwide first.

PROJECT DESCRIPTION

Within the framework of this project, the present contract covered the following two phases :

- Preliminary project : General studies concerning the drillings, well completion, structures and surface installations.
- Project : Detail engineering of the pilot project - consultation with suppliers - orders and requests for tender for the design and construction.

STATE OF PROJECT

The preliminary project phase took place from December 1980 to the second half of 1981 and the project phase ended in June 1983 with the beginning of the initial series of drillings.

RESULTS AND APPLICATIONS

The following work was carried out :

- elaboration of all the detailed programmes needed for the drilling campaign,
- study of specific cements, acidifiable plugging agents, "high temperature" wellheads and a special sliding joint for the tubing of the injection wells,
- a special drilling rig for drilling inclined wells from the surface at an angle of up to 30° was designed, built and adjusted so as to obtain adequate spacing of the wells, in view of the shallow depth of the reservoir,
- study and follow up of the execution of the structures and decks of a drilling and production platform and a utilities platform,
- design and construction of the following equipment :
    - long-stroke reciprocating pumps for the inclined production wells,
    - various types of heat-exchangers,
    - seawater desalination installation,
    - steam generation installation.

On completion of this project, the work is continuing under EEC/GERTH contract n° 05.42/82 : tests.

Title: Nitrogen injection in North Sea reservoirs	Project No.: 05.34/82
Contractor: Britoil PLC Address: 150 St Vincent St., UK - Glasgow G2 Technical director (or person to contact for further information): Mr Taylor	Telephone No.: 041-204 2566 Telex: 777633

Aim of the Project:

To evaluate whether it would be technically feasible and economically viable to enhance oil recovery in certain North Sea reservoirs by injecting nitrogen.

Project Description:

After some preliminary feasibility studies have been carried out a detailed reservoir simulation model will be developed. The detailed simulation study will serve two purposes:

- to ascertain whether nitrogen injection is likely to prove satisfactory and a field pilot test would be worthwhile;
- to check whether natural gas could be used as a cost-saving substitute for nitrogen in the pilot test.

State of Project:

The preliminary feasibility studies have been completed and work has started on developing the Eastern Fault Block model.

Results and applications:

Results of preliminary studies indicated that nitrogen injection offshore could be technically feasible but that economic considerations might favour a number of injection projects sharing a central source of nitrogen, possibly from onshore. They also indicated the need for more detailed reservoir data, hence the plan to develop a detailed, full-field, reservoir simulation model.

If results of project justify it, the study will then be continued, first as a field pilot test and then as a final, definitive technical and economic evaluation of nitrogen injection. Quantitative and economic checks will be made at all relevant key stages to assess whether to continue the work.

Title: Improved oil recovery by means of horizontal drill holes in the Rospo Mare field(offshore Italy - Adriatic sea)	Project No.: 05.36/82
Contractor: Elf Italiana SpA Address: Largo Lorenzo Mossa, 8 Via Aurelia, 619 - I - 00165 Roma Technical director (or person to contact for further information): Mr Jean Paul Demesy	Telephone No.: 39 6 63890 1 or 39 6 63890 323 Telex: 614273 ELFIM I

Aim of the Project:

Try to improve drainage and oil recovery of Rospo Mare field by horizontal well technique because this heavy and viscous oil field presents some risks of early breakthrough with fast water invasion of production wells.

Project Description:

1st stage: including a horizontal well, Rospo 6, to the long time production test planned in Project No. 05.24/81 with 2 wells: Rospo 4 (vertical) and Rospo 5 (directional) both with partial penetration in the reservoir.
2nd stage: interpretation of results obtained on Rospo 6.

State of Project:

The horizontal well production test started in October 1982 and is still continuing (the end of the project itself occurred at the end of 1983). Result interpretation was carried out during first half of 1984.

Results and applications:

- The horizontal well could be pumped at a high rate.
- Breakthrough occurred after a cumulative production of 70,000 m3.
- Hydrated production was not possible because of lack of platform facilities.
- An initial critical rate of 480 m3/d with reversibility of the water coning was evidenced: this critical rate then declines as water contact goes up in the drainage area of the well.
- Result interpretation has consisted in history matching using a simple mathematic formula in order to foresee total oil which can be produced by a horizontal well like Rospo 6.

  The horizon well technique looks promising or an eventual development of this oil field, but it is necessary to realise a production module composed of "n" horizontal wells and to prove its economy.

Title: Recovery of hydrocarbons from bituminous rocks	Project No.: 05.37/82
Contractor: Ammonia Casale Address: Via Lampedusa 13 I - 20141 Milano Technical director (or person to contact for further information): A. Antonini	Telephone No.: (02) 84401 Telex: 321203

Aim of the Project:

To develop a retorting process for the extraction of hydrocarbons from bituminous rocks and/or sands.

Project Description:

The new process to be developed is characterized by the way in which retorting heat is supplied in a special reactor where no intermediates are used (whether gas or inert solid).

The main phases of the project are:

1. Basic study (identification of the necessary chemical physical parameters, optimization of the pilot plant capacity, planning of laboratory tests, preliminary financial calculation)
2. Construction of a pilot plant
3. Laboratory and pilot plant tests.

If the results of Phase 3 are positive, a semi-industrial plant (500 to 1 000 tpd) could be envisaged in a later stage.

State of Project:

The work has started on Phase 1 and the following activities have been performed:

- analysis of existing retorting processes
- definition of the proposed Ammonia Casale process flowsheet and main units
- study of a static, bench scale, pilot plant and its construction
- analysis of laboratory experiment data
- continuous tests on a rotary drum.

Results and applications:

A patent application concerning the new retorting process was drawn up and filed. The tests on a rotary drum have led to a definition of the criteria for the design of a pilot plant.

Title: Examination of the interaction between steam and the distillates of heavy crude in enhanced oil recovery	Project No: 05.38/82
Contractor: University of Technology, Delft Address: Julianalaan 134 NL - Delft Technical director (or person to contact for further information): J. Bruining	Telephone No.: 784 378 Telex: 38151

Aim of the Project:

To enhance the recovery efficiency of heavy oil by the addition of small amounts of its distillables to the injected steam.

Project Description:

Small amounts of distillable oil added to the steam in an oil reservoir are capable of enhancing the recovery efficiency which may approach 100% in the steamed-out region. Without this addition, the comparable efficiency is 80%. Although the steam can only reach part of the reservoir, the new method would mean an improvement in the overall recovery efficiency of a few percent. Ideas are being worked out with regard to high pressure (100 bar) tube experiments and a tentative design for a field test will be made.

State of Project:

The high pressure equipment has been designed and is at present being constructed. A great number of atmospheric experiments and also visual experiments have been made. A preliminary tentative design for a field test has been made.

Results and applications:

The experimental data obtained at present show that for very small amounts of distillable oil added to the steam, the recovery efficiency stays below 100% due to the unstable condensation behaviour of the distillable oil near the steam condensation front.

Phase behaviour of the distillable oil-water system shows that liquid distillables injected in the well will evaporate at the expense of some steam condensation.

Title: CO_2 injection and recovery of crude in the DE4 Pisticci reservoir	Project No.: 05.40/82
Contractor: AGIP SpA Address: San Donato Milanese CP 12069 I - 20120 Milano Technical director (or person to contact for further information): F. Malerba/T. Tassini	Telephone No.: (02) 52 023 227 Telex: 310246

Aim of the Project:

Scope of the project is to ascertain through laboratory research and a field pilot test, whether a huff-and-puff process using supercritical carbon dioxide as "solvent" is able to massively improve well productivity in an existing heavy oil field, which has been exploited for some twenty years at a very low oil production rate.

Project Description:

Four main phases:

- reinterpretation of all existing logs, well and production data etc.; study of reservoir oil and of its mixtures with carbon dioxide; mathematical simulation
- work-over of pilot well
- huff-and-puff with CO_2 for three cycles (200 tonnes), all relevant parameters will be measured and recorded
- evaluation of results

State of Project:

Phase I has been completed. Phase II is progressing.

Results and applications:

The cartography of the reservoir surface (scale 1:10 000) and core analyses have been completed. A test period of 6 months on the wells showed that well PI 13 was the most suitable for the huff-and-puff process. PVT analyses showed that CO_2 saturation (30.94% molar) improves the characteristics of Pisticci oil. Investigation is being conducted about the use of state equation simulator. For the interpretation of the data which will result from the huff-and-puff test, a radial bi-dimensional compositional model has been studied.

Title: Electro-dispersion feasibility	Project No.: 05.41/82
Contractor: GERTH/SYMINEX Address: (Syminex) 2, Boulevard de l'Océan F - 13275 Marseille Technical director (or person to contact for further information): A.J. Kermabon	Telephone No.: (91) 73 90 03 Telex: 400563

Aim of the Project:

This project had three objectives:

1) laboratory experiments to determine the factors involved in the action of DC current on rock samples saturated with water and oil
2) general technological studies (surface and wells)
3) feasibility studies of a pilot test on Eschau field (Alsace).

Project Description:

Laboratory tests were made with Fontainebleau sandstones and Eschau limestones.
Technological preliminary studies included:
- heating effects of current in the well
- well equipment:
  study of different solutions for current transport
  the Anode technology
  well completion
- surface equipment:
  current generator/protection equipment.

Modelling of Eschau field (hydraulic/electrical) were made.
Site studies on Eschau were made.

State of Project:

The project was completed with positive results. However, this specific application on Eschau was considered to be relatively difficult due to the particular geometry of the reservoir and the small diameter of the casing (4" 1/2): at the end of 1982, the pilot test on Eschau field was abandoned. Another pilot test on Lugos field is considered.

Results and applications:

This work showed the technical feasibility of the process and outlined solutions for the method and equipment to be used. However, to go to the next step of realizing a pilot test would require a considerable extension of technical studies to achieve detailed specifications of equipment.

Title : EMERAUDE PILOT PROJECT : TESTING PHASE	Project n° :05.42/82
Contractor : GERTH Address : 4, av. de Bois Préau 92500 RUEIL-MALMAISON FRANCE Technical director (or person to contact for further information) : Mr Gilbert BLU or Mr Bernard MERCIER	Telephone n° : (1) 749.02.14 ext. 2288 or 2747 Telex : 203 050 F

AIM OF THE PROJECT

This Franco-Italian project involves a pilot project for injecting steam into Emeraude field (People's Republic of the Congo). It represents a worldwide first offshore.

The test installations were designed under contract TH 05.33/81 "Emeraude steam pilot project : feasibility". The purpose of this contract was to carry out preliminary testing of injection of steam into the reservoir.

PROJECT DESCRIPTION

- Tests of the well injectivity and productivity.
- Tests of the installations before startup.
- Pumped primary production.
- Starting of steam injection.

15 wells were to have been drilled under the initial programme, but it was considered wiser in view of the difficulties encountered during drilling (losses - plugs) to drill only three wells in the initial stage.

Water and steam injectivity tests were carried out in order to test the cement seal and passages between the layer and the hole. Furthermore, these were taken advantage of to carry out interference tests with neighbouring platforms.

The results of this test campaign were considered satisfactory. However, they led to a modification to the drilling programmes and to the abandon of the two observation wells and the two "huff and puff" wells.

Reservoir studies

Based on analysis of the data (cores, drilling logs and well tests) obtained, these concern the preparation of an image of the reservoirs with a view to future simulations (laboratory studies on flows in a porous medium).

STATE OF THE PROJECT

The project continues with the second drilling campaign for 6 other wells, that will be carried out in 1984.

RESULTS AND APPLICATIONS

The results will be known after completing steam injections, which will start at the end of 1984.

Title : TESTING OF NEW PRELIMINARY TECHNOLOGIES ON HEAVY OIL FIELDS	Project n° : 05.43/82
Contractor : GERTH - ASVAHL Address : 4, av. de Bois Préau 92500 RUEIL-MALMAISON FRANCE Technical director (or person to contact for further information) : Mr. Jean François LE PAGE	Telephone n° : (1) 749.02.14 ext. 2288 or 2747 Telex : 203 050 F

AIM OF THE PROJECT

The overall project covered the construction and testing of demonstration units combined together onto Solaize platform, in order to study the various methods of valorization of heavy oils on the production field. The first part of the project was a feasibility study, whilst the second covered the construction proper of the platform. The third part of the overall project is the subject of the present contract, and is not yet complete.

PROJECT DESCRIPTION

The programme comprises two distinct phases of work :

- startup of the platform units, implying a number of prior technological tests ; the main units involved are desalination, atmospheric distillation, vacuum distillation, de-asphalting, hydroprocessing, visbreaking and hydrovisbreaking ; some of these units comprise improvements and technological innovations, the soundness of which should be checked
- adaptation of the hydroprocessing unit so as to render it capable of operation in a boiling bed and a back-flow mobile bed.

STATE OF THE PROJECT

As regards startup, the various units and all the associated secondary units were loaded with oil, apart from the visbreaking and hydrovisbreaking units. These various tests enabled the correct operation of this equipment to be checked, or in the event of the contrary the necessary modifications to be made before proceeding to the test proper. This initial phase of the contract will be completed in October 1984.

As regards adaptation of the hydroprocessing unit, the process file is complete, as also are the layout studies and basic engineering. Tests of various types of European valves to inject the fresh catalyst and remove the spent catalyst are now nearing completion. Automatic monitoring and control of this catalyst circulation loop has been studied and defined. The detail engineering studies will be complete in December 1984.

RESULTS AND APPLICATIONS

The results will be known at the end of the project.

<table>
<tr><td>Title : INJECTION OF CO2 INTO
PECORADE RESERVOIR</td><td>Project n° : 05.44/82</td></tr>
<tr><td>Contractor : GERTH

Address : 4, av. de Bois Préau
92500 RUEIL-MALMAISON
FRANCE

Technical director (or person to contact for further information) :
Mr Gilbert BLU or Mr Bernard MERCIER</td><td>Telephone n° : (1) 749.02.14
ext. 2288 or 2747

Telex : 203 050 F</td></tr>
</table>

AIM OF THE PROJECT

This project concerns the implementation at PECORADE reservoir of pilot injection of CO_2 to maintain the reservoir pressure above the critical pressure whilst at the same time enhancing oil recovery in a part of the reservoir difficult to exploit using conventional methods. It follows a feasibility study performed under contract TH 05.29/81.

PROJECT DESCRIPTION

The CO_2 is extracted from Lacq gas, liquefied and stored at - 20° C and at pressure of 20 bars. It contains less than 1 000 ppm of H_2S and is then carried by tank trucks to the injection site 65 kilometres away, where it is again stored into 175 m^3 containers. Injection takes place in liquid form at an average daily rate of 60 tons of CO_2 for 420 bars at the injection wellhead (PCE 13). The conditions in the three production wells (PCE 04, PCE 22 and PCE 23) about 200 metres away from the injection wells, are followed by gauges and weekly samples of effluents, so as to be able to follow the action of the CO_2 in the reservoir.

STATE OF THE PROJECT

The operations of starting up the surface installations and pilot project wells took place from 1st July 1982 to 15th March 1983.

Injection of CO_2 was started in mid-March 1983 and ended on 7th March 1984, injecting a total of 10,700 tons of CO_2.

RESULTS AND APPLICATIONS

The technical phase of the project (production, liquefaction, transport and injection of CO_2) took place satisfactorily, except for a failure of the CO_2 injection pumps, which resulted in a lengthy interruption of injection from 18th October 1983 to 6th February 1984.

The very poor petrophysical characteristics of the reservoir have required a high injection pressure in the range of 420 bars at the wellhead (PCE 13), to maintain an injection of an acceptable level (60 tons per days). These pressure conditions have resulted in fracturation inside the reservoir at the injection well, whereas initially injection was to take place at pore communication level. The result of this fracturation has been very quick breakthrough of the CO_2 into the production wells, so that the surrounding matrix was only affected little by the injection.

ENVIRONMENTAL INFLUENCE ON OFFSHORE STRUCTURES

Title: Feasibility study on a submerged wave dampener	Project No.: 06.05/76
Contractor: Bertin & Cie Address: BP 3, F - 783737 Plaisir CEDEX Technical director (or person to contact for further information): M. P. Facon	Telephone No.: (3) 056 25 00 Telex: 696231 F

Aim of the Project:

To verify the feasibility of a submerged wave dampener for protecting offshore equipment.

Project Description:

The dampener consists of two caissons equipped with flexible valves opening one way, either inwards or outwards, so as to establish water circulation between one caisson and another, dissipating energy by turbulence.

State of Project:

Feasibility study.

Results and applications:

Tank tests on a model have shown that wave attenuation over 60% can be achieved.

Various applications can be envisaged:

- Protection of installation sites
- Protection of loading buoys
- Stabilisation or protection of semisubmersible platforms
- Electricity production by using a turbine.

Title: New hydraulic offshore pile hammer	Project No.: 06.07/78
Contractor: BSP International Foundations Ltd Address: Claydon, Ipswich UK - Suffolk IP6 OJD Techncial director (or person to contact for further information): Mr Murray/ Mr Storey	Telephone No.: (0473) 830 431 Telex: 98115

Aim of the Project:

BSP, recognizing the problems of extracting oil and gas under conditions of increasing water depth, have developed a range of hydraulic hammers which are capable of operating underwater.

Project Description:

The development was centred on hydraulic actuators of 10 t and 20 t capacity. The simple design of the hammer chassis and the modular construction of the actuators allows any number of them to be used according to the size of hammer required. The actuators are easily replaced in the event of a failure so that spare units obviate the need for a complete stand-by hammer.

The hammer does not operate in an air bell, so small air hoses to the surface are unnecessary and the impact cap, which provides for the most efficient transmission of impact force through to the pile, does not need a separate hydraulic supply as is the case with nitrogen buffer cushions.

Results and applications:

A pair of 10 t actuators mounted on a hammer have operated satisfactorily in 120 m water depth and an electrical control system has been developed and proven at the same depth.

Other developments include a slender hammer which is able to pass through pile guides, removing the need for pile followers as is the case with above-water driving.

Title: Underwater pile driving test of an off-shore pile driver hydraulically operated by an underwater powerpack	Project No.: 06.08/80
Contractor: Bomag-Menck GmbH Address: Werner von Siemenstrasse 2 D - 2086 Ellerau Technical director (or person to contact for further information): Messrs Hahlbrock/Kühn	Telephone No.: 04106/7002-0 Telex: 213294

Aim of the Project:

To prove safe operation of a new developed pile driver as mentioned above.

Project Description:

Testing procedure grouped into three stages:

- Operational test with underwater powerpack ashore at the Menck plant in Elelrau;

- Submerging test with underwater powerpack at a harbour site;

- Underwater pile driving test with the pile driver, incorporating the underwater powerpack, in course of a jacket installation offshore.

State of Project:

Finished in 1982.

Results and applications:

Tests have been completed successfully. A clear effect of rationalisation has been proved and by this a new pile driving equipment could be introduced into the offshore market.

Title: Mooring system for floating production platform	Project No.: 06.09/80
Contractor: Société européenne de Propulsion - Division Propulsion à Poudre et Composites Address: BP 37 F - 33165 St Médard en Jalles CEDEX Technical director (or person to contact for further information): Mr Yves Appell	Telephone No.: (56) 34 84 90 Telex: SEP 560678

Aim of the Project:

The project was related to the development of an articulated system for the mooring on the sea bed of a positive oscillating tower. The objective was to develop a system using rubber-metal laminated bearings. This system avoids wear and maintenance problems which are usually met with the metal friction solutions.

Project Description:

The developed system comprises a spherical main bearing consisting of six 60° sectors and an axisymmetrical secondary bearing capable of taking up incidental downward load and enabling preloading of the system. The maximum loading conditions specified were: upward load: 2 200 tonnes; horizontal load: 750 tonnes; swivel angle: 14 degrees.

State of Project:

Design, scale 1 60° sector and scale 1/5 test model have been achieved.

Results and applications:

A scale 1 60° sector has been successfully manufactured, proving the system feasibility.

On the tests carried out on a scale 1/5 model, conditions equivalent to the specifications have been achieved. No failure or damage has been observed during the test of this model.

Title: Adaptation of a gravity foundation system to removable production platforms	Project No.: 06.10/80
Contractor: Sea Tank Co. Address: Sea Tank Co./C.G. Doris 58 A, rue du Dessous des Berges F- 75013 Paris Technical director (or person to contact for further information): M. Vaché (C.G. Doris)	Telephone No.: (1) 584 11 64 Telex: 270263

Aim of the Project:

To adapt a gravity foundation system to removable production platforms.

Project Description:

Breakout forces required to retrieve objects embedded in clay may be of the magnitude of the bearing capacity of the soil, but the experimental conditions are very different from offshore platform conditions. A laboratory model test on clay samples under a hydrostatic pressure of 1 MPa was carried out in order to get a better understanding of the phenomena. The project includes:

- selection of soil characteristics
- complete bibliographical study
- changes in soil characteristics under platform loadings
- experimental study of suction at the soil-structure interface; investigation of means to accelerate suction dissipation
- theoretical approach of the phenomena
- design of suitable foundation systems and devices in order to cope with retrieval requirements.

State of Project:

The related tests and studies were completed in October 1981.

Results and applications:

Model tests have demonstrated the feasibility of platform retrieval by minimizing the bond forces through drainage at the soil-structure interface. Some systems and devices to achieve the drainage conditions have been studied, and have to be integrated into the concrete structure at the design stage. Although this project has not yet been the subject of any application, there is a potential for its partial application in the future.

Title: Investigation into the behaviour of piles under tensile loads	Project No.: 06.12/80
Contractor: Taylor Woodrow Construction Ltd. Address: 345 Ruislip Rd, Southhall UK - Middlesex UB1 2QX Technical director (or person to contact for further information): Dr R.D. Browne	Telephone No.: 01 578 2366 Telex: 24428 TAWOOD G

Aim of the Project:

The purpose of the programme is to aid the development of rational approaches to the design of anchorages and foundations for offshore structures, based upon driven piles acting in tension. This work is particularly relevant to the safe and economic design/development of the new generation of floating/compliant offshore structures for hydrocarbons exploiration.

Project Description:

The major part of the programme will be directed towards a co-ordinated series of piling trials to support the review and development of analytical techniques. The first phase will be single pile tests on field scale. These tests will be conducted at one or more selected land based sites having soil properties in the range compatible with typical North Sea sediments. The output from this experimental programme will be used in the validation of the analytical techniques being studied in parallel. The second phase of the work will be based on group piling tests. The basic knowledge of pile performance obtained in the first phase will provide the base parameters for an extension of the field work, single and group tests at model scale in the laboratory, and a limited number of small scale tests in a centrifuge to study the global performance of anchorage/foundation systems.

State of Project:

Not commenced.

Results and applications:

–

Title: Testing of novel offshore pile hammers	Project No.: 06.13/81
Contractor: BSP International Foundations Ltd. Address: Claydon, Ipswich UK - Suffolk IP6 0JD Technical director (or person to contact for further information): Mr R M Elliott	Telephone No.: 0473 830431 Telex: 98115

Aim of the Project:

To demonstrate the ability of novel piling hammers to operate over sustained periods reliably.

Project Description:

Tests were carried out on 48 and 12 tonne metre hydraulic hammers and also a 12 tonne metre slim hydraulic hammer. These tests were carried out both above and below water and also inside a pile.

State of Project:

Complete.

Results and applications:

The tests were successfully completed and a number of pile anchors driven. However commercial application of the hammers has yet to be achieved.

Title: A field investigation into the performance of a piled foundation system for an offshore oil production platform	Project No.: 06.14/81
Contractor: The British Petroleum Co. PLC Address: Britannic House Moor Lane, UK - London EC27 9BU Technical director (or person to contact for further information): Mr W J Rigden, Manager Civil and Geotechnical Branch	Telephone No.: 01-920 8264 Telex: 888811

Aim of the Project:

The aim of the project is to obtain full scale data on the behaviour of the foundation of an offshore oil production platform.

Project Description:

BP Petroleum Development Ltd. have installed a production platform in 186 m of water at Magnus Field in Block 211/12 of the UK Sector, North Sea.

The project consists in instrumenting the lower section of one of the four legs of the Magnus Platform and piles supporting that leg. The behaviour of the instrumented members is recorded and a comparison made between measured pile loads and those calculated by conventional methods.

State of Project:

The instrumentation and data acquisition unit (DAU) has been commissioned. Data was recorded on a temporary unit from March 1982. The permanent DAU was operational from December 1983 and is currently recording information.

Results and applications:

Raw data from the permanent DAU is being analysed. A full report should be available by November 1985.

<table>
<tr><td>Title: Arge cryotrans</td><td>Project No.: 06.16/82</td></tr>
<tr><td>Contractor:
Salzgitter AG
Address:
Postfach 411129
D - 3320 Salzgitter 41
Technical director (or person to contact for further information):</td><td>Telephone No.:
(05341) 21-1
Telex:
954481</td></tr>
</table>

Aim of the Project:

To check by means of large-scale tests the economic and technical viability of offshore transfer systems for cryogenic liquids.

Project Description:

The project is a continuation of the first (engineering) phase which produced the overall design and prepared the detailed testing program. In the present project, the system components will be manufactured and checked for their proper functioning.

The testing plant consists of a steel tower with a tower head which permits full 360 degree horizontal rotation, and a 40m long boom. The main phases of the construction work will be:

- tower
- tower head with boom, control and monitoring system
- hose transfer system with control and monitoring system
- tensioning system for hose and articulated transfer system
- construction and assembly of the mooring system
- preparation for modifying the transport vessel
- final acceptance and operational tests.

State of Advancement:

Although a contract signature was made in February 1983, the project has not been implemented. Discussions are being held with possible Norwegian participants.

AUXILIARY SHIPS
AND SUBMERSIBLES

Title: Remote-controlled subsea handling vehicle	Project No.: 13/75
Contractor: Winn Technology Address: Kilbrittain EI - Co. Cork Technical director (or person to contact for further information): R. Winn	Telephone No.: Bandon 49601 Telex: 8443

Aim of the Project:

Development of a remote-controlled subsea handling vehicle

State of Project:

The project was completed in August 1979 as per the contract.

Results and applications:

The project culminated in the production of an operational machine capable of:

- manoeuvring in any direction on the seabed under control from a console on the surface, navigation being accomplished by resolving direction and distance information received from the on-board gyro compass and wheel rotation; the resulting components are indicated on a chart at the control console
- manipulating objects up to 250 lbs in weight at the extremity of the arm which is fitted with a manipulating claw, developed under contract 7.88/76; the claw, which is not removable, can in turn pick up additional devices developed under contract 7.88/76
- performing operations such as valve turning and the unscrewing of nuts, bolts and clevis pins.

The launch and recovery gear unit is skid-mounted, complete with its power generator and umbilical cable reel. The console is mounted separately.

Title: Equipment of a special ship for soil investigation	Project No.: 07.01/76
Contractor: Preussag and Partner Address: Postfach 4829, Arndtstrasse 1, D - 3000 Hannover 1 Technical director (or person to contact for further information): Dr Amedick	Telphone No.: (0511) 123 25 31 Telex: 922851

Aim of the Project:

To equip a special ship for soil investigation.

Project Description:

Components of the total system:

- drilling equipment
- in situ investigating equipment (Penetro meter)
- geophysical equipment
- Vibrocone equipment
- ship "Berliner Tor".

Drilling device and coring equipment for unconsolidated sediments were developed. Experience in solid rock could not yet be gained. The Penetro meter was carefully tested and modified.

The geophysical equipment is a combination of Precision Echo Sounder, Side Scan Sonar, Boomer, Sparker and a digital system. The system has been tested and is operational. The development of a processing programme has been finished.

Two vibration devices ("Kieler Hammer" and "Senkowitch") have been tested on the coring device winch that will be installed and are operational.

The ship "Berliner Tor": tests, sea trials and mooring tests have demonstrated that the ship is very suitable for use in the southern part of the North Sea.

Title: Soil investigation in the North Sea	Project No.: 07.02/76
Contractor: Fugro Cesco Address: Veurse Achterweg 6, Postbus 41 NL - 2260 AA Leidschendam Technical director (or person to contact for further information): A.J.A. Van Overeem	Telephone No.: (070) 209 250 Telex: 31010

Aim of the Project:

Soil investigation in the North Sea.

Project Description:

Anchor Packer

After the sea trials in May, where the basic feasibility of offshore use was proved, a lot of additional effort had to be spent on existing drilling and testing systems. As usual some additional investments had to be made to carry out the necessary field test to prove the technical feasibility and show the marketability to clients. This latter process is expected to take some two years more before a final evaluation can be made about the success/failure of this part of the research project.

Seabed Jack

In November the seabed jack was tested offshore. This first use showed, besides some minor technical problems, the overall technical viability of the concept. These problems concerned mainly the underwater full opening of the clamp. The tests showed that a lot of simplification in operation and design will be required before the tool can eventually be offered economically. The result of this part of the programme consists mainly of an increased level of know-how for this type of system and its impact on standard operations.

Re-entry system

This more simple approach to understand problems on rigging and seabed stability had its influence by improving the design of the seabed jack. No action was taken to develop this particular feature further for better performance, as it proved to be basically awkward.

Results and applications:

Overall, in this project a much better understanding was gained on vertical drillstring control. One of the three prototype tools developed showed enough promise to develop further operationally to try to find an economical use for it.

Title: Development of submersible and support vessel. Subproject 1: Submersible Mermaid VI	Project No.: 07.05/76
Contractor: Bruker Physik AG (now Bruker Meerestechnik GmbH) Address: Wikingerstrasse 13 D - 7500 Karlsruhe 21 Technical director (or person to contact for further information): Dipl. Ing. J Haas	Telephone No.: (0721) 5967-182 Telex: 7825656

Aim of the Project:

Development of a deep-diving submersible for offshore underwater work down to 600m depth with diver lock-out and rescue capability.

Project Description:

Design, construction and testing of a work submersible for 600m depth including specially developed components such as manipulator arms with exchangeable tools, heat pump diver heating system, exchangeable lock-out trunk/rescue drome, hydraulic underwater thrusters, high-pressure drain pumps etc.

State of Project:

Project resulted in commercial sales and was successfully completed in 1980.

Results and applications:

Two submersibles type Mermaid VI for 600m diving depth with diver lock-out and rescue (dry transfer) capability were delivered and are now in use in the Mediterranean sea.

Title Development of a submersible and support vessel. Subproject 2: Tauchkatamaran	Project No.: 07.05/76
Contractor: Bruker Physik AG (now Bruker Meerestechnik GmbH) Address: Wikingerstrasse 13 D - 7500 Karlsruhe 21 Technical director (or person to contact for further information): Dipl. Ing. J. Haas	Telephone No.: (0721) 5161-182 Telex: 7825256

Aim of the Project:

It was the aim of the project to abolish problems connected with conventional handling techniques during launch and retrieval of submersibles in bad weather by introducing a new kind of support vessel.

Project Description:

Various possibilities of launch and recovery from a catamaran vessel, capable of self-diving, were investigated. Consequent pursuit of this idea led to the Subcat concept, a submersible catamaran vessel suited as mothership for a small submersible but also as a fully autonomous underwater work system.

State of Project:

Feasibility and design studies were carried out and demonstrated the attractiveness of the concept. Due to the future-oriented technology and because of budgetary reasons, the project could not be completed at the time.

Results and applications:

Idea found considerable attention. Project could not be realised for commercial and technical reasons (lack of surface, independent power supply). No application so far.

Title: A submerged vehicle tool system	Project No.: 07.08/76
Contractor: Winn Technology Address: Kilbrittain EI - Co. Cork Technical director (or person to contact for further information): R. Winn	Telephone No.: Bandon 49601 Telex: 8443

Aim of the Project:

Development of a submerged vehicle tool system.

Project Description:

1) A claw device has been developed which carries out a number of the requirements without the employment of special tools. This device has been fitted to the machine developed under contract 13/75 and has proved capable and adaptable in many situations. It is capable of continuous rotation and can therefore be used to operate rotating tools without the addition of an extra power unit. A special grip receptacle is provided to enable tools to be gripped concentrically and thereby rotated.
2) Screwdriver tools, drills, spanners and saws can be provided with this grip, but it should be noted that on many occasions the claw jaws are capable of gripping nuts, bolts, valves etc. without additional tooling.
3) The rotary brush tool is held in the claw device and can undertake various tasks through the employment of various sizes and grades of bristle. Brush diameters up to 22" have been employed.
4) An ultrasonic surveying head has been developed, incorporating a liner array of transducers enabling objects to be detected. This can be mounted on the rotating portion of the wrist to provide the necessary scanning in azimuth and bearing.
5) A submerged stud welder has been demonstrated wherin the energy imparted to the weld produces a pressure rise in the molten metal exceeding ambient sea pressure, hence producing metallurgically uncontaminated weld.
6) An impact driver has been developed: results from this were disappointing due to limitation of terminal velocity achieved from internal mass. An electro magnetic device is under investigation, using a pulse power supply developed for the welding device described above.
7) Seismic investigation of subsea surfaces has proved possible using the impact driver in conjunction with the ultrasonic surveying head.

Results and applications:

Operation of these devices is readily possible in the company's in-house test tank facility where the necessary operator training has been carried out.

Title: Deep water trenching vehicle TM 402	Project No.: 07.13/77
Contractor: Tecnomare SpA Address: S. Marco, 2091 I - 30124 Venezia Technical director (or person to contact for further information): Paolo Vielmo	Telephone No.: 708622 Telex: 410484

Aim of the Project:

Design, construction and sea test of an underwater vehicle for the trenching of flexible pipes and electrical cables already laid on the sea bottom, in order to guarantee their protection against possible damage.

Project Description:

The underwater unmanned tracked vehicle is remote-controlled from the surface, moves on the sea bottom self-guiding along the pipe or cable, digs a trench with vertical walls by means of a chain device in the soil with a compressive strength of up to 150 kg/cm^2.

The main capacities of the vehicle are:

- water depth	up to 160 m
- trench depth	1.5 m
- cable pipe diameter	5÷30 cm
- max trenching speed	400 m/h

State of Project:

Completed.

Results and applications:

The vehicle has been successfully tested at sea by trenching a bundle of flexible pipes and an electrical cable. In order to satisfy the market requirements, the experience obtained during the project development has been utilised for the design of two vehicles, the first of which was finalised for the burying of flowlines (300 mm O.D. max), the second for the burying of electrical cables only.

Title: Development of an underwater machine capable of trenching in rocky sea bottoms to bury sea lines	Project No.: 07.14/77
Contractor: SAIPEM Address: S. Donato Milanese I - 20100 Milano Technical director (or person to contact for further information): Mr Casagrande/Mr Gioielli	Telephone No.: (02) 535 31 Telex: 31246

Aim of the Project:

The research programme to develop the pipeline trenching machine will be carried out with the intention of studying and verifying the operation of individual components before proceeding with the development of the prototype.

Project Description:

After having assembled in our logistical centre at Cortemaggiore the necessary testing equipment, we have tested individual components such as evacuation pumps and samples of our cutting wheel with various cutters in different configurations. We have succeeded in verifying the validity and efficiency of various rotating systems and hydraulic bearings. Having reached these conclusions, we looked at the resources of various trustworthy contractors and commissioned a feasibility study.

Because of a change in market conditions, this project was stopped in December 1980.

Title: Medium and long range positioning system (MEROPS)	Project No.: 01.04/76 07.15/77
Contractor: SERCEL Address: Avenue Bel Air, BP 64 F - 44471 Carquefou Cedex Technical director (or person to be contacted for further information): G. Nard	Telephone No.: (40) 30 11 81 Telex: 710695

Aim of the Project:

1. Extension of the power budget of the SYLEDIS system in order to reach distances of 80 to 200 miles while keeping a high degree of accuracy (10 to 30 metres)
2. Development of real time on board distance data processing to introduce reliable distance errors correction and Kalman filtering through the use of microprocessor techniques (UCM)

Project Description:

Design of special UHF antenna stacked arrays and space diversity disposition. Development of a high power booster amplifier and improved distance processing techniques. Complements to the "Unité de Calcul microprogrammée" (UCM), a microprocessor-based on-board computer, developed for the SYLEDIS system to implement distance errors filtering, coordinate transforms and Kalman filtering of the position.

State of the Project:

The project was completed in 1976 and 1977 and was the subject of the report "Exact Medium Range (400 kilometres) Radiolocation of Vessels: the MEROPS system (medium range positioning system)" at the EEC Luxembourg Symposium, 19 April 1979.

Results and applications:

The extension of power budget of the SYLEDIS system has been proved effective both for long range extension and reliability and accuracy of distance measurements. The use of space diversity aerials and proper setting up on antennae in conjunction with booster amplifiers has been particularly efficient. The implementation and use of Kalman filtering in UCM was also successful.

From 1976 to 1983 inclusive, the following equipment has been manufactured and set into operation for oil prospection all over the world:

195 - booster amplifiers and filters
120 - antenna switches for space diversity purpose
140 - UCM microprocessor - controlled unit.

Title: Navigation methods for the prospection of hydrocarbon off the Continental shelf	Project No.:07.16/77
Contractor: Prakla-Seismos GMBH Address: Buchholzer Str. 100, D - 3000 Hannover 51 Technical director (or person to contact for further information): Dipl. Ing. B.E. Gerlach	Telephone No.: (0511) 642-0 Telex: 922847, 922419, 912250

Aim of the Project:

Development of an integrated navigation system for prospection of hydrocarbon off the Continental shelf with sophisticated filter technique to achieve required accuracy in deep water also.

Project Description:

The integrated navigation system with A PDP 11/34 process control computer and microprocessors as intelligent interface units combines doppler sonar velocity inputs and gyro heading inputs, measurements from Loran C system or other radio navigation systems and information from the Transit satellite system optimally by a filter algorithm with the aim to minimize the user state estimation uncertainties.

State of Project:

The development phase under the contract of the Commission of European Communities lasted from October 1977 to June 1980.

Results and applications:

The hydrocarbon exploration required absolute accuracy of 150 m CEP; relative accuracy of 100 m CEP and relative sequential accuracy of 15 m CEP has been reached with the threefold integration off the Continental shelf. Beside the navigational requirements the system realises a very simple, self-explanatory interactive man/machine communication by structured dialogue. The system has now been installed on two seismic vessels, the "SV Explora" and "SV Prospekta", and on one oceanographic vessel, the "DPFVS Polarstern".

Title : IMPROVEMENT OF THE TECHNOLOGY OF DEEP SEA EXPLORATION	Project n° : 07.29/78
Contractor : GERTH Address : 4, av. de Bois Préau 92500 RUEIL-MALMAISON FRANCE Technical director (or person to contact for further information) : Mr Gilbert BLU or Mr Bernard MERCIER	Telephone n° : (1) 749.02.14 ext. 2288 or 2747 Telex : 203 050 F

This project includes 2 parts :

Development of a sediment corer

A preliminary study performed in 1979 led to the concept of a truly fixed piston tubular corer enabling undisturbed cores to be sampled by the penetration of the corer tube.

In order to test this concept, a small dimension (diameter of cores 54 mm) prototype corer C1 was built and applied in the Mediterranean in 1980, at the same time as a conventional piston corer (a Kullenberg corer). The comparative results obtained led in 1981-1982 to the construction of a large size (core diameter : 110 mm) prototype corer C2.

The tests performed in 1983 in the Mediterranean in varying depths of water of from 50 to 1 250 metres with a 20 metre corer tube enabled excellent quality cores to be sampled endowed with a filling ratio of 90 to 100 %, which was far above those obtained with conventional oceanography corers. These tests enabled the reliability of the methods of application of a device with an overall length of 25 to 30 metres and a weight varying from 5 to 10 tons to be verified.

Improvement to the CYANA diving vessel

The CYANA diving vessel is a small manned submarine capable of attaining a depth of 3 000 metres. It is used by CNEXO for a variety of missions in deep water and the work carried out under the project was aimed at improving the technical means used during these missions.

A corer enabling samples of hard rock 100 mm in length and 20 mm in diameter was built and tested at sea down to depths of 300 metres. The results obtained showed that considerable work will have to be performed before this equipment can be considered as operational.

An acoustic navigating system enabling the CYANA to obtain a fix on the bottom independently was developed and tested. After modifications and adjustment, it will be usable for specific missions in great depths of water.

A measurement acquisition system was developed for the CYANA and is now operational.

Title: Design, construction and tests of a long-range manned submarine	Project No.: 07.34/80
Contractor: S.S.O.S. Address: Via della Scafa 19 I - 00054 Roma-Fiumicino Technical director (or person to contact for further information): Mr Liuzzi	Telephone No.: (06) 645 30 41 Telex: 611156

Aim of the Project:

Design, construction and experimentation of a self-contained manned submarine having wide range, long duration and great operating depth.

Project Description:

Design, construction and fitting of an established operational depth of a submarine, built with toroidal structure - navigational crew at atmospheric pressure, operational crew in hyperbaric or atmospheric pressure.

State of Project:

Pressure hull in advanced construction, mathematical model for structural analysis underway.

Propulsion system is undergoing tests to select and build some specific components of the closed circuit. Most of the equipment and various components involved in the final fitting have been defined, identified and are under acquisition. Meanwhile all the drawings and designs for the definitive project are on their way to completion.

Results and applications:

Very important results have been acquired both while pressure hull was under construction and during tests on power plant. Great flexibility is kept in the solutions during the various stages of the project, so that today it is possible to assign different tasks to the submarine:

- saturation lock-in - lock-out submarine
- hyperbaric rescue boat
- atmospheric rescue submarine
- survey.

Title: Remote-operated servicing and maintenance system for subsea oil equipment	Project No.: 07.35/80
Contractor: Alsthom Atlantique ACB Address: Prairie au Duc F - 44040 Nantes Cédex 2 Technical director (or person to contact for further information): M. de Vaulx	Telephone No.: (40) 47 32 31 Telex: 710960

Aim of the Project:

The project aim is to develop tools and equipment necessary to improve performance and make the remote-operated subsea vehicles more competitive.

Project Description:

The project comprises four phases:

- phase A: system and equipment design
- phase B: developing of specific equipment
- phace C: system integration
- phase D: sea trial

State of Project:

Phase A of the project has been completed and a synthesis report sent to the Commission.

Results and applications:

The study enabled all the conditions of submarine remote-control operation to be established and solutions proposed to improve the mechanical tools and automatisation of this activity. The proposed solutions cover critical engineering and technology:

- sea handling systems
- autonomy and manipulation
- automatic steering of the vehicle.

Profitable applications of this study are in the field of submarine oil production in areas of diverless submarine operations.

Title: Personnel Transfer Unit	Project No.: 07.36/80
Contractor: Ateliers et Chantiers de Bretagne Address: Prairie au Duc CEDEX n° 2, F - 44040 NANTES CEDEX Technical director (or person to contact for further information): J.P. Manesse	Telephone No.: (40) 47 31 31 Telex: 710960

Aim of the Project:

Further to the developments of the oil industry at sea, and in particular the North Sea, the problem of transferring personnel onto the platforms is becoming a more definite and demanding problem.

The current methods are not entirely satisfactory for security reasons (basket) and costly (helicopter). ACB therefore decided to study, construct and make operational a Personnel Transfer Unit. This unit, which is complementary to offshore transportation by boat to platforms, consists of a rigid basket handled by the cranes available on the platform.

Project Description:

The proposed unit is made up a cabin comprising two mechanical units, with two winch reels. The superior reel is used for the cable which supports the cabin, the extremity of this cable being attached to the crane hook. The inferior reel is used for the rope which is attached to the deck of the boat permitting the guiding of the cabin during transfer manoeuvres.

Each of these reels is capable, when the other is locked with a dog, of compensating the heaving motion of the boat.

State of Project:

The development programme of the Personnel Transfer Unit comprises of three phases: Phase 1: Studies; Phase 2: Construction - shop tests; Phase 3: Sea trials.

Results and applications:

Due to various mechanical incidents during the prototype workshop tests, ACB decided to stop the project in its original concept, as its reliability was insufficient.

Title: DAVID - A versatile multipurpose submersible for remote control or diver assisted performance	Project Nos.: 07.22/78 07.33/79 07.43/81
Contractor: ZF-HERION Systemtechnik GmbH Address: Postfach 2168, D -7012 Fellbach Technical director (or person to contact for further information): Mr Klaus Wiemer	Telephone No.: 0711/507-351 Telex: 7 254 733 zfhs d

Aim of the Project:
To specify, design, manufacture and test a submersible system for universal application in the construction, inspection, maintenance and repair sectors of the offshore industry.

Project Description:
The component parts of the DAVID system are the submersible vehicle, the deployment device, the umbilical handling system, the surface control station and the power supply unit.

The submersible unit can either be remotely operated from the surface control station or locally by the diver working on site. The vehicle has highly developed control and navigation system and carries both equipment for remote controlled inspection duties as well as power tools and test equipment for use by the diver. Other facilities include an arrangement for docking onto a subsea structure and a moveable platform for use by the diver.

State of Project:
Specification, design and manufacture of prototype system have been completed. First phase of testing was successfully completed in Bergen, Norway and system is now on board a diving support vessel undergoing final evaluation trials in North Sea.

Results and applications:
The tests in Norway were carried out during May 84 under the guidance of Det Norske Veritas. The system generally performed as specified and intended, and a statement to this effect was issued by DNV. More results will become available as the system progresses through final trials during 2nd half of 1984.

Applications for the system are in:
- exploration drilling
- Subsea inspection; Observation; cleaning; Non-destructive testing; Potential measurement; Riser and pipe inspection
- Construction, Platform installation; subsea completion systems; Loading buoys; Pipeline surveying; Pipelaying; Pipeline completion; Pipeline tie-ins
- Maintenance and Repair; Cleaning; Welding; Anode changing.
  Other applications can be found in military operations (e.g. torpedo recovery) and marine salvage.

Title: Development of a remote-controlled submersible	Project No.: 07.45/82
Contractor: ZF-Herion-Systemtechnik GmbH Address: Postfach 2168 D - 7012 Fellbach Technical director (or person to contact for further information): Klaus Wiemer	Telephone No.: (0711) 507 351 Telex: 7 254733

Aim of the Project:

The project is referred to as the "MARS" project (maintenance and repair submersible) and has the objective of developing a system for carrying out inspections, maintenance and repair on sub-sea equipment and structures. An operational underwater service depth of 1 500m is anticipated.

Project Description:

The "MARS" work unit will be remotely controlled either from the surface or from a second manned atmospheric submersible, operating locally with it. In addition, this work unit will have the following features:

- high available power for heavy work
- possibility of attaching the submersible to an underwater structure
- possibility of bringing a range of various tools by means of a tool arm, a tool magazine and remotely operated tool changing system
- a manipulator arm.

The "MARS" project is divided into 3 phases, the first only of which is the subject of the EEC support. The main stages of this first phase are:

- task definition
- market research
- feasibility study
- preliminary design
- preliminary research.

State of Project:

The project has commenced. Work has started on the initial specification which will define where research and development should be invested to achieve the desired end result.

<table>
<tr><td>Title:

Development of a deep water tethered manned submersible</td><td>Project No.: 07.46/82</td></tr>
<tr><td>Contractor:
T.E. Associates Ltd.
Offshore Systems Engineering Ltd.

Address:
Elm House, Clanwilliam Court
Lower Mount Street
EI - Dublin 2

Technical director (or person to contact for further information):
T.C. Earls</td><td>Telephone No.:
(01) 685 222

Telex:
30488</td></tr>
</table>

Aim of the Project:

Design, manufacture and testing of a tethered one-man submersible capable of performing sophisticated tasks on hydrocarbon drilling/production installations in water depths up to 1 500 metres.

Project Description:

Three areas of technical innovation related to hull design, underwater deployment system, and the manipulators. A further area of innovative design related to the adaption of the vehicle for optional remote control.

State of Project:

Component design about 85% complete. Final prototype design and assembly 50%. Initial trials scheduled for late 1984.

Results and applications:

Problem areas relating to the hull, tether deployment system and manipulators were defined and resolved. In the case of the manipulators the second stage of a planned three-stage development has produced a viable tool which may be operated by a single joystick control and uses sea water as its hydraulic fluid.

Initial tests of a dual control system have been successfully completed. This enables the vehicle to be remote-controlled from the surface for performance of simple tasks or where hazardous conditions might prevail.

Title: Vehicle for inspection and maintenance of marine structures in deep water	Project No.: 07.48/82
Contractor: Tecnomare SpA Address: S. Marco, 2091 I - 30124 Venezia Technical director (or person to contact for further information): M. Mazzon	Telephone No.: 708 622 Telex: 410484

Aim of the Project:

To study and develop an unmanned untethered underwater vehicle capable of performing visual inspection, surface cleaning and NDT controls on marine structures in deep waters.

Project Description:

The project development is composed of four phases:

- analysis of the mission, issue of the system specification, preliminary design
- engineering of the system and execution of laboratory test of critical components and subsystems
- construction and shop tests
- sea trials.

The vehicle is composed of a vessel containing the control system and the energy source, the propulsion system, a device to connect the vehicle to the structure, and the manipulator system.

State of Project:

Preparation of system specifications (first phase).

Results and applications:

The vehicle mission, the estimated power consumption and the energy requirements have been defined. The vehicle general configuration has been defined and critical areas (energy sources, cleaning method etc.) analysed.

PIPELINE LAYING

Title: Diverless flowline and pipeline connections	Project No.: 6/75
Contractor: Comex Seal (GERTH) Address: Comex SA F - 13275 Marseille Cédex 2 Technical director (or person to contact for further information):	Telephone No.: (91) 41 01 70 Telex: 410985

Aim of the Project:

- to define the techniques and equipment which enable diverless flowline and pipeline connections to be made
- to demonstrate the feasibility of the so defined methods.

Project Description:

As it is an interface problem, multiple or single connections between sea bed located structures in great water depths, made without diver intervention, can only be considered in particular cases. Nevertheless, enough specific solutions can be defined by the study of the following problems:

- single line at arrival or departure from a structure, when bottom shifting and pulling-in are possible
- single line associated to a cluster: (internally) connection between tress and gathering manifold; (externally) connection between gathering manifold and production support.

Results and applications:

1. Single line connections
   - Beginning of laying (first connection)
     The barge "Anguille" was moored above the template. The laying tool with the hauling cable and the reversing pulley was hooked on the template. The cable, secured to the end of the 2 inch pipe, guided it to the reception frame of the template. A connector carried out the diverless connection.
   - End of the laying
     The 2 inch flexible COFLEXIP pipe was laid down according to the COFLEXIP method. At the free end, floats are installed along the line. Then a cable is secured to the pulling head of the flexible line. The cable, guided by the pulley of the special pulling tool hooked on the template, guided the flexible line down to it.

2. Single line connection associated to a cluster
   - internal connections: preliminary studies permitted a remote machine to be designed, capable of intervening in a subsea structure
   - external connections: laying and connections of a 6 inch pipe and electrical cables. The line was assembled on shore. Then, floating 2 m above the sea bed, the pipe was towed to the site. Difficulties were encountered (positioning, adjustment of the overlength collars for the floats), but the laying succeeded. Cables were also laid, but the connectors were not tight enough, resulting in this part of the project being unsuccessful.

Title: Laying tests in the Straits of Messina	Project No.: 20/75
Contractor: SNAM SpA Address: 20097 S Donato Milanese I - 20100 Milano Technical director (or person to contact for further information): M. Cicarelli	Telephone No.: (02) 520 59 14 Telex: 310246

Aim of the Project:

The scope of the tests was to verify the feasibility of laying down a pipeline in deep waters up to 360 mt, with heavy environmental conditions such as difficult geomorphological strucutres and strong, quickly variable sea currents.

Project Description:

The aim of the tests was to experiment with new measuring and control systems, to verify the accuracy of the mathematical models used in the computer software, to specify the technical limits of the laying procedure and the possibility of coordinating laying operations by submarine assistance. For these tests a pipe with the following characteristics was used:

- diameter 10" (27.3 cm)
- thickness 15.88 mm
- material according to API STD 5LX-X52
- polythene external coating

State of Project:

Complete.

Results and applications:

The main conclusion was that the techniques employed allowed the pipe to be laid to a maximum depth of 360 mt.

The conditions of the sealine were satisfactory, both as regards its integrity and its static support on the bottom, and this was largely proved by the pressure test and by the many inspections made by submarine.

Finally the equipment used was shown to be suitable and convenient for the designed applications.

Title: Laying tests in the Sicily Channel	Project No.: 21/75
Contractor: SNAM SpA Address: 20097 S. Donato Milanese I - 20100 Milano Technical director (or person to contact for further information): M. Cicarelli	Telephone No.: (02) 520 59 14 Telex: 310246

Aim of the Project:

The scope of the tests was to verify the feasibility of laying down a pipeline in deep waters (up to 550 mt), focusing on the relevant laying procedure and acquiring at the same time a wide range of data.

Project Description:

In order to have the largest possible amount of information, it was decided to lay two different sections of pipeline (12" and 16") 3.2 km and 3.5 km long respectively.

Both the sections have been laid down at the maximum depth and in such a way as to cross flat and uneven areas.

The other main characteristics of the pipes used are:

- thickness (16") 17.48 mm and (12") 15.88 mm
- materials according to API STD 5LX-X52 and to special specifications
- external coating (for testing only) selected in the following types: polythene, neoprene (2 kinds), epossidic resin.

State of Project:

Complete.

Results and applications:

The most important conclusion is that the techniques employed enabled the pipes to be laid to a maximum depth of 550 mt.

The conditions of the sealine was satisfactory, both as regards its integrity and its static support on the sea bottom, and this was largely proved by the pressure test and by the many inspections made by submarine. On the other hand, it is appropriate to point out that the lay barge used proved to be, as foreseen, particularly sensitive to the meteorological conditions, also taking into account the necessary routing orientation practically perpendicular to the direction of the most dangerous storms.

This, although problems were caused in one aspect, enabled experience and data to be gained to improve all the equipment employed.

Title: New technology for pipelaying at sea	Project No.: TH 09.06/76
Contractor: Bouygues Offshore Address: La Boursidière - RN 186 F - 92357 Le Plessis Robinson Cedex Technical director (or person to contact for further information): Louis Auperin	Telephone No.: (1) 630 21 98 Telex: 204420

Aim of the Project:

To define and analyse all the components of a new technology for pipe-laying in great water depth. This technology uses a flexible laying unit with the capability of a very long and flexible stinger.

Project Description:

Design and calculation of the flexible laying unit:

- static bi- and tri-dimensional computer programmes for the S curve calculation
- dynamic S curve computer programme. Engineering and tests of ballasting system
- design of flexible laying unit including the pipelaying support device
- design of the S curve control system:
  - . study of pipelaying configurations in great water depths (1 000 - 2 000 m)
  - . model tests: dynamic behaviour of the FLU during laying
- behaviour of a laying barge during laying - dynamic positioning of the barge.

State of Project:

After the steps described above had been carried out, the project was stopped at the end of 1978.

Results and applications:

The general feasibility of this new technology of laying pipes in great water depths was proved. Laying tests would be necessary to complete the study.

Title : DEEP SEA CONNECTING TECHNIQUES AND PIPELINES	Project n° : 09.07/77
Contractor : GERTH Address : 4, av. de Bois Préau 92500 RUEIL-MALMAISON FRANCE Technical director (or person to contact for further information) : Mr Gilbert BLU or Mr Bernard MERCIER	Telephone n° : (1) 749.02.14 ext. 2288 or 2747 Telex : 203 050 F

AIM OF THE PROJECT

The objectives of the project were first to examine the possibilities of the RAT method (end-connection of a section of towed-up pipeline) under tension, on the surface, and then to study the J-configuration laying method, so as develop a subsea mechanical connector for pipelines and a method of repairing pipelines under water.

STATE OF PROJECT

The project started on 1st March 1977 and ended on 31st December 1979.

DESCRIPTION AND RESULTS OF THE PROJECTS

Optimization of the RAT method

The study showed that the field of application of this method, tested within the framework of contract 18/75, was in actual fact limited to the development of the satellite fields of the major oil-producing fields.

J-configuration laying method.

Major general research on the method, the dynamic behaviour of the pipe during laying and the connection techniques, were followed by tests intended to prove the feasibility of components (joints, ramps...) or welding systems. This led to the choice of a connection by electron-beam welding and the drafting of preliminary project of a welding machine. The work was continued under contract 09.19/80.

Pipeline repair method

General study of the problems set by repairing pipelines has revealed the importance of careful preparation of the end, carried out automatically in great depths of water. Feasibility studies have proved that the tools were capable of operating underwater, and a preliminary project of a repair system was developed.
The work was continued under contract 09.19/79.

Repair with mechanical connectors

The studies and tests carried out in this area can only be considered as an approach to mechanical connection. They have demonstrated the feasibility of mechanical connection by means of an air-tight metal-to-metal seal against a formed collar.
The work was continued under contract 10.21/79.

Title: Pipelaying method in very deep waters with mechanical connections	Project No.: 09.17/79
Contractor: Tecnomare SpA Address: S. Marco 2091, I - 30123 Venezia Technical director (or person to contact for further information): E. Palla	Telephone No.: 708622 Telex: 410 484 MAREVE I

Aim of the Project:

Design, construction and shop test of two single station pipeline junction methods (mechanical connector and TIG automatic welding plant) suitable for saline J-laying procedure.

Project Description:

The project is subdivided into the following main phases:

- preliminary design of the welding machine, the mechanical connector and the laying system;
- engineering of the welding machine plant and of the mechanical connector with subsequent construction and shop tests of the prototypes.

The welding machine plant is composed of three tungsten inert gas torches mounted on a frame and rotating around the pipeline joint during welding. An automatic computerised system checks and controls continuously the welding parameters. The mechanical joint is of threaded type and the sealing is provided by metal to metal plastic contact.

State of Project:

Welding machine and mechanical joint are under construction.

Results and applications:

Presently the main results achieved consist of the design of the two different solutions for the connection of pipeline sections. The construction and test of the two prototypes are under way.

Title: Automatic processing of side scan sonar records	Project No.: 09.18/79
Contractor: Sesam (France) Address: 132 ave de Villeneuve St Georges F - 94600 Choisy-le-Roi Technical director (or person to contact for further information): Mr Philippe Gaudillère or Mrs B Robert	Telephone No.: (33.1) 890 82 44 Telex: 202 268 SESAM F

Aim of the Project:

Automatic processing of side scan sonar pictures to obtain more rapid and more precise results.

Project Description:

- Photograph of the sonar picture taken through a cross ruled screen, detail by detail;
- identification and interpretation of the various forms based on the different types of shades (64 types of grey) as well as on the geometry of the shapes;
- restitution: automatic correction of perspective of the picture.

State of Project:

Feasibility study of picture analysis using records in our hands now. Problems of interpretation still have to be studied.

Results and applications:

The feasibility study gave good results: the dynamic of the picture is sufficient. The rapidity of the processing (5 to 30 seconds with 35 x 35 cm picture) is satisfactory together with the resolution which is 0.7 millimetres.

It is possible to identify peculiar forms such as pipeline, wreck, or the totality of the picture with all recorded details (sea bottom features, geological information, etc.).

Title : DEEP SEA REPAIR	Project n° : 09.19/79
Contractor : GERTH/COMEX/ALSTHOM-ATLANTIQUE Address : 4, av. de Bois Préau 92500 RUEIL-MALMAISON FRANCE Technical director (or person to contact for further information) : Mr Gilbert BLU or Mr Bernard MERCIER	Telephone n° : (1) 749.02.14 ext. 2288 or 2747 Telex : 203 050 F

AIM OF THE PROJECT

The objective of this project is to design, build and test in the workshop and at shallow depths an automatic system for preparation of the ends of pipelines, controlled from the surface, in order to connect up pipelines on the sea bottom.

PROJECT DESCRIPTION

The system has been designed to carry out the following tasks, down to depths of 1000 metres :

- removal of the concrete,
- removal of the anti-corrosion coating,
- clean cutoff of the end of the tube,
- cleaning of the interior,
- machining of the longitudinal welding bead on the tube.

It comprises a structure that lifts the end of the pipeline and attaches to it, together with a removable table on which the tools for machining the ends of the pipeline are laid.

The prototype robot has been built for a useful depth of 300 metres and has carried out tasks 2 to 4 above.

The system is modular. No divers are used. The handling operations are performed without guide lines, the "package" being positioned by acoustic and video systems. The robot is fully remote-controlled from the surface. It is assisted by a remote-controlled submarine (ROV) for its operational, observation and emergency functions.

STATE OF PROJECT

The various items of equipment of the prototype system have been built. Workshop testing of the complete system is now nearing completion. Shallow depth trials will be starting in June 1984 at Saint-Nazaire.

RESULTS AND APPLICATIONS

Great depth trials will start in the Summer of 1984. They will be performed under contract 10.42/83.

Use of this system is of interest for the connecting devices of pipelines at great depth and in particular :

- the atmospheric welding connection procedure (Weldap method),
- the procedure for connecting by means of mechanical couplers,
- the hyperbar welding method (inasmuch as its depth of use can be increased).

Title: Rip-Rap protection of underwater pipelines by Velpo vehicle	Project No.: 09.21/79
Contractor: Velpo Group - Composed of: C.G. Doris - OTP - CEA - CFD Address: C.G. Doris 58 A, rue du Dessous des Berges F - 75013 Paris Technical director (or person to contact for further information): M. Jean Martin	Telephone No.: (1) 584 11 64 Telex: 270 263 F

Aim of the Project:

Rip-Rap protection is one of the most efficient means ensuring a good protection of the submerged pipelines which cannot be buried, by giving good weight stabilisation, reducing free spans and protecting the pipe against anchors and falling objects. The aim of the project was to improve the methods used for dumping stones over pipelines, in order to reduce the costs of such Rip-Rap protection.

Project Description:

The system consists of an underwater vehicle, which can be described as a mobile hopper, capable to travel over the submerged pipeline. The hopper is large enough to accommodate the movements of the feeding pipe of a stone dumping vessel, dynamically positioned above the vehicle. As the vehicle is accurately positioned above the pipeline, all the stones dumped from the vessel are collected and end up nicely laid above the pipeline to be covered. The system is rated for operation down to 300 meter water depth.

State of Project:

The design of the vehicle including all the positioning systems have been completed, and a prototype has been tested underwater.

Results and applications:

The tests on the prototype have been successful and the system is now ready for full scale operation.

Title: Development of an ultrasonic riser pig	Project No.: 09.18/80
Contractor: Röntgen Technische Dienst BV Address: Delftweg 144 NL - 3046 NC Rotterdam Technical director (or person to contact for further information): J.A. de Raad Ing.	Telephone No.: (010) 150 200 ext. 291 Telex: 23366

Aim of the Project:

To develop a tool to inspect riser pipes under operating conditions. The tool should be suitable to enter the riser pipe to 300 metres from the launch-retrieve trap and use an umbilical/rescue cable.

Project Description:

The integrity of riser pipes is of paramount importance. To prove its condition the ultrasonic method was selected to measure pipe wall thickness and quantify corrosion, if present. Inspection from the outside is often impossible, therefore tools are required which can enter the riser pipe system, preferably under operating conditions.

Typical riser operation conditions are pressures up to 150 Bar and oil or gas velocities up to 5 m/sec.

State of Project:

The contract has been completed.
A new project for free swimming tools is under consideration.

Results and applications:

A special cable and winch unit were built and worked satisfactorily. Ultrasonic probes were selected and a prototype ultrasonic system constructed. Prior to a high pressure system a tool was constructed to inspect open pipelines to a distance of 500 metres using the constructed ultrasonic device. The field test showed its good performance and justification of the project. Attention was also paid to high pressure free swimming tools, as such the RTD-Caliper was designed.

Title : J-CURVE PIPE LAYING ELECTRON BEAM WELDING	Project n° : 09.19/80
Contractor : GERTH Address : 4, av. de Bois Préau 92500 RUEIL-MALMAISON FRANCE Technical director (or person to contact for further information) : Mr Gilbert BLU or Mr Bernard MERCIER	Telephone n° : (1) 749.02.14 ext. 2288 or 2747 Telex : 203 050 F

AIM OF PROJECT

The J curve pipe laying method, associated with the electron beam welding, seems to be the most attractive method to lay thick pipes of high steel grades in great water depths.

The object of this project consisted in building a complete prototype machine for welding pipelines with electron beam, and testing this machine at the workshop on a test bench that simulates the movements of the laying barge in order to prepare for the qualification by the certification authorities.

PROJECT DESCRIPTION

The work programme included the successive phases :

- the engineering study of the laying ramp, mainly the inclined mast and loading arm,
- the study of the welding machine including :
  - a mating system enabling to position both extremities of the pipes to be welded, using soft tongs
  - a prototype welding machine
  - a retractable vacuum seal with important deflection
  - an internal clamp
- the "prequalification of the welding process"
- the manufacturing and the testing of the units of the welding machine
- the test of this equipment on an oscillating test platform.

STATE OF PROJECT

The project started on 15th January 1979 and ended on 31st December 1981.

RESULTS AND APPLICATIONS

Tests were performed at vertical position then at inclined position, with or without roll.

The tests were carried out with series of 3 pipes, making it possible to execute two welds without interrupting the roll motion on 24" pipes of X 65 and X 100 steel grades.

Tests made it possible to conclude that the envisaged rythm of a weld every 15 minutes could easily be obtained.

A reliability study, showed that the probability of achieving a weld cycle without incident represented 99,5 % (this figure can be compared to that of a radar on board a ship).

TRANSPORT

Title: Deep steel pipeline	Project No.: 18/75
Contractor: GERTH Address: 4, ave de Bois Préau F - 92500 Rueil-Malmaison Technical director (or person to contact for further information):	Telephone No.: (1) 749 02 14 ext. 2288 or 2747 Telex: 203050

Aim of the Project:

The purpose of the project was to solve the problems set by laying, repairing and using subsea pipelines with diameters of up to 40", under difficult sea conditions and in depths of up to 500m. This project had the following two main objectives:
- to improve existing techniques and means and if possible extend their limits
- depending on the operational or economic limits of these techniques, to study and develop new methods.

State of the Project:

The project began on 1 January 1974 and ended in February 1978.

Project Description and Results:

The development and application of the dynamic computation programme turned out to be a very delicate matter and this programme cannot be considered as operational.

The studies performed on anchoring systems and the equipment of the laying barges enabled the depth limits to be set at 300 to 400m for a 30" pipeline and about 800m for a 12" pipeline.

No applications were found for the anti-heave system for handling heavy packages, owing to its complexity and size.

Three new techniques underwent testing in the North Sea:
- A test of towing a 16" pipeline equipped with buoys and 1 000m in length in Moray Firth in December 1975. This test was performed in turn on the surface and then sub-surface (-15m), in winds of force 8 to 10, and proved the feasibility of this method, with sub-surface towing coming out ahead.
- A test on laying with the RAT method (end-connection under tension of a section of pipeline towed to site) near Stavanger in June 1977. A false manoeuvre of the tug caused by bad weather broke the towed section.
- A test of welding a pipeline at great depth using the WELDAP method. The tests took place in February 1978 in Erdfjord near Stavanger in 265m water depth. A manual weld was made on a 20" diamer tube inside a chamber maintained at atmospheric pressure.

Title: General study and field testing of deep water submarine pipeline	Project No.: 19/75
Contractor: Tecnomare SpA Address: S. Marco 2091 I - 30124 Venezia Technical director (or person to contact for further information): Mr Rodighiero	Telephone No.: 70822 Telex: 41484

Aim of the Project:

The scope of the project was the study and field testing of the laying of pipelines in deep waters.

Project Description:

The project was constituted by the definition and the development of the technology to be followed in order to achieve pipe laying capability in water depths up to 1 000m, with pipe sizes ranging from 6" to 48".

State of Project:

Complete.

Results and applications:

The main technical results were the following:

a) computer procedures: the set-up calculation methods allow all the principal conditions the laying equipment may meet to be analysed

b) model tests: model tests have been performed on the different components (barge, stinger, mooring line, pipe etc.) and on the complete laying system for analysing the behaviour of the stinger pipe system and the moored lay barge

c) field tests: the experimental data collected during the sea trials carried out by SNAM in the Sicily Channel have been analysed and compared with the theoretical results derived from the computer procedures; both the model and field tests confirmed the validity of the computer programmes

d) study of new techniques and equipment: many techniques and equipment necessary to solve specific problems encountered during pipe-laying operations in deep waters have been identified and studied.

Title: Laying of a 12" flexible at 550m	Project No.: 10.04/76
Contractor: COFLEXIP Address: 23, Avenue de Neuilly F - 75116 Paris Technical director (or person to contact for further information): J. Laurent	Telephone No.: 747 711 42 Telex: 610302

Aim of the Project:
To lay a 12" flexible at 550m depth.

Project Description:
At an earlier stage, we have manufactured a prototype and simulated on land the working conditions of the flexible:
- hydrostatic crushing due to external pressure
- static axial traction
- dynamic movements.

a) To determine the resistance to collapse (55 bar) with the outer layer pierced, the following material was used:
   - compression chamber
   - pressure delivery system
   - reduction in volume measuring system
   - measuring thermometer.

   Before being put into the chamber, the flexible is filled with water and connected to the measuring system. While the pressure in the box is increasing, the flexible, decreasing in volume, expels some of the water which it contains. This water, collected in a graduated cylinder, gives the amount of the volume decrease.

b) To verify the tensile strength (160 t), the force corresponding to a flexible submerged at 550 m, the following material was used:
   - a tensile machine
   - measuring apparatus.

   The sample was held at one end on a fixed point of the bank by a ferrule, the other extremity was attached to the jack of the traction block.

c) A study of the behaviour of a flexible laid on the bottom but not touching the soil and therefore exposed to cyclical forces by the current was examined with the aid of a fatigue machine.

   The bending is given by the inclination of the jaws which hold the extremities of the flexible, rotation being implicit in this arrangement.

Title: Offshore LNG transfer system	Project No.: 10.06/76
Contractor: David Brown-Vosper (Offshore) Ltd Address: Graphic House, Castle Street, Porchester UK - Hampshire P016 9PH Technical director (or person to contact for further information): J A Guest	Telephone No.: Cosham 83331 Telex: 86156

Aim of the Project:
To develop an offshore transfer system to handle liquid natural gas between floating structures containing liquefaction and storage facilities and LNG carriers to the shore terminals. The design will consider North Sea environmental conditions.

Project Description:
The project comprises the development of an offshore transfer system including a new cryogenic swivel with special safety features, suitable for operation in a marine environment. The development was planned in four major phases:

Phase I: design, develop and construct the LNG swivel carry out swivel test program
Phase II: design, develop and prove the LNG transfer system
Phase III: production of the LNG transfer system
Phase IV: effect sea trials of LNG transfer system between floating structure and LNG carrier

State of Project:
The contract was prematurely concluded at the end of 1979 due to:
- slower than anticipated interest of oil companies in the procurement of offshore LNG transfer systems, which significantly affected the hardware development proposal
- additional costs incurred in the development of a leak-proof cryogenic swivel.

Results and applications:
Phase I:
A design study of 6 swivel types was undertaken, the cylindrical roller swivel being considered most suitable. However, tests proved that this type could not be selected due to difficulty with bearing in cryogenic. Further tests concentrated on the ball bearing type swivel. The single ball track capsule design swivel was found to be successful. An efficient cryogenic swivel sealing system was developed.
Phase II:
Investigations were conducted into the transfer system in the following categories: submerged or floating flexible hose; above water rigid pipe; above water flexible hose. The contractor studied in more detail three systems:
- flexible hose twin catenary transfer system
- hybrid single catenary hose system
- articulated rigid pipe system/
The systems are based on proven engineering design and technology, except for cryogenic hoses.

Title: Waste heat recovery system Thermobloc (pipeline booster station)	Project No.: 10.12/77
Contractor: Borsig GmbH Address: Berliner Str. 27-37 D - 1000 Berlin 27 Technical director (or person to contact for further information): H Mattes Dr W Malewski Dep BS	Telephone No.: (030) 43011 Telex: 018204-0

Aim of the Project:

Realization of a waste heat recovery system using gas turbine exhaust.

Project Description:

The prototype plant, using conventional components (boiler, turbine, pump, piping system), produces mechanical energy from the waste gas of a gas turbine via a closed process (Rankine cycle), with ammonia as a working fluid under operating conditions above and below the critical point. The system extends and supplements an existing natural gas compressor station driving an additional gas compressor. The system was designed to be operated unattended from remote locations.

State of Project:

The project was realised on a turn-key basis in Canada, province of Alberta. The waste heat recovery system was started up, commissioned and tested in 1982 and 1983. The system was accepted by the client.

Results and applications:

The main point of investigation was the decomposition of ammonia. Acceptable rates can be obtained if the ammonia temperature is kept below 300°. The waste heat recovery system as designed and performed is capable of operating under Arctic conditions (ambient temperatures +25 to -40°C) on almost predicted preformance data.

Title: Cryogenic flexible pipe	Project No.: 10.14/78
Contractor: Coflexip Address: 23, ave de Neuilly F - 75116 Paris Technical director (or person to contact for further information): M. Hamadi DRIDI	Telephone No.: (1) 747 11 42 Telex: 610 302 F

Aim of the Project:

To work out a new product found upon the Coflexip flexible pipes principle, which could convey fluids at very low temperatures, particularly LNG products.

Project Description:

A. General studies on materials, flexibles, end fittings

B. Creation of testing means

C. Manufacturing of flexibles

D. Tests on the flexibles

State of Project:

Phases A, B and C are completely finished. Phase D results are expected in December 1984.

Results and applications:

-

Title: Development of high pressure flexible pipes	Project No.: 10.16/78
Contractor: Dunlop Ltd; Oil & Marine Div. Address: Moody lane, Pyewipe Industrial Estate Grimsby, UK - South Humberside DN31 2SP Technical director (or person to contact for further information): Mr A D Griffiths, Project Manager	Telephone No.: 0472 59281 Telex: 52184

Aim of the Project:

To design and develop high pressure flexible pipes to carry sweet and sour, diphasic, crude oil at temperatures up to 130°.

Project Description:

To develop flexible pipes which will satisfy the following:

a) survive multiple rapid decompressions

b) operate at working pressures up to 10,000 psi

c) resist hydrogen sulphide and carbon dioxide.

State of Project:

New plant and equipment is in advanced stages of installation. Project will be completed by the end of 1984, pipes up to 150 ft. long with bore size 2" to 12" will be available.

Results and applications:

All objectives are met. Dunlop Limited is now able to market the following:

a) Light weight rotary drilling hose
b) Kill and choke hoses
c) Fire resistant covers (optional)
d) Spool pieces and jumpers
e) Vibrator hoses
f) Flexible connectors for rigid pipe systems

Title: Offshore loading system for LNG/LPG and other refrigerated fluids	Project No.: 10.17/78
Contractor: Salzgitter AG Address: Postfach 41 11 29 D - 3320 Salzgitter 41 Technical director (or person to contact for further information): R. Holekamp	Telephone No.: (05341) 21-3962 Telex: 954481

Aim of the Project:

Aim of the project was to develop an efficient articulated offshore loading system for LNG and LPG which is reliable even under rough sea conditions. This system will permit profitable utilisation of small gas fields and associated gas occurring when oil is extracted.

Project Description:

The developed system for transferring liquefied gas even under extreme environmental conditions from a liquefaction platform to an ocean carrier consists of the following main components:

- loading arm (all metal arms composed of rigid pipe sections and swivel joints)
- ship coupling system
- supporting structure
- tensioning system
- computerised monitoring and control system

State of Project:

The project was terminated in June 1982, after the planned programme was completed.

Results and application:

Extensive studies, computations, computer simulations and tests have shown that the developed system is a feasible and efficient solution. This was also demonstrated by motion simulation tests in 1 to 10 scale. The articulated offshore loading system was precertificated by Det Norske Veritas (Oslo).

Nevertheless, it seems necessary to prove the economy of the system by large-scale tests under offshore conditions as a basic requirement of an economic application.

Title: Deep water pipeline repair system	Project No.: 10.20/79
Contractor: SNAM SpA Address: P.O.Box 12060, I - 20120 Milan Technical director (or person to contact for further information): Ing. Mario Baudino	Telephone No.: (02) 520 5914 Telex: 310246 ENI

Aim of the Project:

To realise an unmanned repair system to repair sealines in deep water.

Project Description:

The system is modular and is composed of the following sub-assemblies:

- one launch and recovery system
- one D.P. thruster and power module
- two H - frames
- one pipe preparation and cutting module
- one end connector and spool replacement module
- one damaged pipe recovery module
- one dredge module
- end connectors and spool piece.

State of Project:

All modules are under construction.

Results and applications:

At the present time the first remarkable result consists in having developed a new metal-to-metal ball end connector, two female mating connectors and slip joint (patented) for sealine repairs.

All these products are easily remotely installable. Applications other than for on-bottom sealines repairs, are foreseen for platform riser connections.

Other applications of the whole system will be evaluated during the test phase scheduled for summer 1985.

Title : REPAIR OF SUBSEA PIPELINES BY MECHANICAL COUPLINGS	Project n° : 10.21/79
Contractor : GERTH Address : 4, av. de Bois Préau 92500 RUEIL-MALMAISON FRANCE Technical director (or person to contact for further information) : Mr Gilbert BLU or Mr Bernard MERCIER	Telephone n° : (1) 749.02.14 ext. 2288 or 2747 Telex : 203 050 F

AIM OF THE PROJECT

The objective of the project is to build a device enabling a damaged section of pipeline lying in very deep water to be replaced automatically.

PROJECT DESCRIPTION

This technique for repairing subsea pipelines enables a deformable U-piece equipped with two mechanical couplers and a metal-to-metal seal to be installed, after forming a collet at each end of the line.

On the ends of a subsea pipeline already prepared, namely by de-concreting, brushing and a making a clean cut, the following operations are required :

. forming a belled collet on each end through the action of a hydraulic punch. The necessary forming pressure is taken up by an elastic ring which attaches itself inside the pipeline. The forming module also has an active centering device to facilitate insertion into the pipeline,
. metrology of the collets, comparing them one with the other, by means of an articulated gauge that can be extended hydraulically under remote-control,
. insertion of a deformable "S-piece" manufactured to suit the gauge and equipped with hydraulically actuated mechanical couplers. The metal-to-metal type seal between the nose of the coupler and the unadjusted collet is brought about by tightening a metal joint until plastic deformation occurs.

STATE OF PROJECT

A 12" prototype coupler has been undergoing endurance tests for over a year on a circuit containing gas with a high acid content.

Prototype equipment with a diameter of 20" has been designed with a view to sea trials. Following manufacture and assembly, all the equipment is now either already acceptance-tested or in the process of being so.

RESULTS AND APPLICATIONS

These will be known only when the tests demonstrating the feasibility of the project have taken place.

Title: Self-destroying instrumented vehicle for the inspection of pipelines	Project No.: 10.26/81
Contractor: SYMINEX Address: 2, Boulevard de l'Océan F - 13275 Marseille Cédex 9 Technical director (or person to contact for further information): M. Fiocchi	Telephone No.: (91) 73 90 03 Telex: 400563

Aim of the Project:

The object of the project is to study construction and testing of a low-cost self-destroying instrumented vehicle destined for the maintenance control of pipelines (corrosion monitoring).

Project Description:

The tool developed by SYMINEX will operate in a 12" diameter pipe in a 10 to 20 km length line. It is made of "syntactic foam" and could be self-destroyed by a pyromechanism if it is jammed for a long time in the pipe. In addition, a magnetic sensor gives information about the corrosion of the pipe. The data acquisition and storage are managed by an on-board microcomputer.

State of Project:

The studies to assess the feasibility of the project are partly achieved. Tests of several parts of the vehicle are in progress.

Results and applications:

The SYMINEX vehicle is designed as an alarm tool and will be able to give general information about the corrosion state along a pipeline. It will enable a regular low-cost curvey of the line, in order to detect corrosion appearance and monitor its evolution. It cancels the risk of a production stoppage due to the jamming of the vehicle.

Title: LNG - LPG Offshore Transfer System	Project No.: 10.28/81
Contractor: FMC Europe SA (on behalf of FMC Europe SA Brussel) Address: Route des Clérimois F - 89102 SENS CEDEX Technical director (or person to contact for further information): M. Gilles GRONEAU	Telephone No.: (86) 65 65 45 Telex: 800477

Aim of the Project:

The aim of the project is to identify the most critical parts of equipment in LNG-LBP transfer system and design this part in order to solve specific offshore problems.

Project Description:

The most critical part identified is the swivel joint likely to be used in all kinds of transfer systems (including flexibles). This part is generally used to cope with combined motion, load and fluid transfer and then is subject to fatigue, wear, thermal, mechanical problems, the first two parameters being specific to the offshore environment.

State of Project:

A solution has been designed specially to solve fatigue and wear problems together with keeping up the standards to be met in classical cryogenic transfer.

This solution makes use of a unique component in this field which is an arrangement of free wheels intended to spread out the wear all over the wear surfaces instead of allowing it to be localised as it is the case in most oscillating devices.

Results and applications:

The prototype is now built and the start up phase of tests completed. These tests are being made at full scale with LNG flow. The prototype will be tested for 10 million cycles with loads and thermal cycles truly simulated.

Title : OFFSHORE LOADING OF LIQUEFIED GASES	Project n° : 10.35/82
Contractor : GERTH /EMH Address : 4, av. de Bois Préau 92500 RUEIL-MALMAISON FRANCE Technical director (or person to contact for further information) : Mr Gilbert BLU or Mr Bernard MERCIER	Telephone n° : (1) 749.02.14 ext. 2288 or 2747 Telex : 203 050 F

AIM OF THE PROJECT

This project is intended to develop an offshore terminal for refrigerated liquefied gases and particularly liquefied petroleum gases (LPG). The terminal envisaged comprises a loading station in the open sea capable either of loading an LPG tanker (product at - 45°C) from a shore storage facility, or the reverse.

PROJECT DESCRIPTION

The project comprises the following subsystems :

- transfer line between shore storage facility and loading station,
- offshore loading station with single mooring point enabling the vessel to tie up whilst remaining free to turn completely around the station so as to lie to the winds and currents,
- transfer line between this single point mooring (SPM) and the ship.

STATE AND RESULTS OF THE PROJECT

After comparing the various possible preliminary projects, the choice and predimensioning of the optimum design for the terminal was made, including floating hoses, with a CALM (catenary anchored leg mooring) buoy, fully motorized and remote-controlled, with a low temperature bi-fluid rotary joint.

A second phase of the project enabled each of these subsystems to be dimensioned exactly and the problems set by the service temperature and offshore environment to be solved.

A third phase of the project now taking place is undertaking the tests needed to qualify certain components and critical materials, namely :

- low temperature static tests for the sealing linings of the bi-fluid rotary joint to be preselected,
- low temperature cyclic fatigue testing (- 48°C) for a prototype of the rotary joint,
- welding tests to select steel grades for GPL pipeline,
- floating hoses adapted to transfer of liquefied gas.

A final phase will be completed in 1984 :

- low temperature testing of a significant length of prototype pipeline,
- qualification by low temperature cyclic tests of the hoses proposed by the manufacturing firms.

A pilot project file will be prepared and will conclude the present project.

Title: Prototype construction and prototype testing of the UW-work and pipeline repair system "Supra"	Project No.: 10.37/82
Contractor: Arge Supra Address: c/o Ocean Consult GmbH Halbmond 30d, D-2058 Lauenburg Technical director (or person to contact for further information): R.D. Klaeke, Ocean Consult GmbH & H. Fiebig, Howaldtswerke-Deutsche Werft AG	Telephone No.: 04153/2314 Telex:

Aim of the Project:

Supra is designed to ensure unlimited working autonomy at sea bottom.

Project Description:

New and versatile type of underwater work system for 420 m water depth and integrated into a catamaran-type floatable vessel. It can be operated either manned and diver-assisted or remotely controlled and unmanned. It is equipped with underwater television and an obstacle avoidance sonar system. For diver-assisted UW-works, conventional offshore saturation diving systems are used.

State of Project:

Supra is expected to be ready for operation in the summer of 1985.

Results and applications:

Supra's main applications are:

1. Gripping and aligning of pipelines by the four integrated alignment frames;
2. Supply of hydraulic and electric power for the machining of pipelines or structures on the sea bottom;
3. Hyperbaric dry welding in the integrated welding habitat;
4. Interconnecting of pipelines by mechanical couplings;
5. Transport of heavy loads to the sea-bottom by a storage platform and platform and underwater handling thereof by integrated underwater cranes;
6. Inspection and service works at the sea bottom, either in the "dry" habitat or with the telescope swivel crane.

Supra's versatile applicability together with the vessel's capability to float and to dive and its independence of large surface barges renders it particularly economic and attractive to operators and companies which only dispose of smaller diving and supply vessels.

12. NATURAL GAS TECHNOLOGY

Title: Cryogenic pipeline	Project No.: 12.03/76
Contractor: OTP Address: 3 & 5, Rue Volta F - 92 Puteaux Technical director (or person to contact for further information): M. Van Tuyen	Telephone No.: 506 2194 Telex:

Aim of the Project:

The objective of this study was to examine the possibility of transporting LNG for long distances. Several aspects have been considered:

- technical and technological problems
- transportation economics
- possibility of application in Community countries
- environmental impact.

Project Description:

Beginning with the existing state of the art and existing materials, it is possible to demonstrate the technical feasibility of an LNG pipeline. Several solutions pertaining to the structure of the line were developed. The economic study, taking into account the actual price of steel or nickle, allowed a comparison between the transportation of natural gas in its gaseous form and its liquid form - in certain cases the transport of LNG can be more economic. This method of transport permits the retrieval of refrigeration at the pipeline terminal.

Cryogenic transport may be the solution to moving large quantities from producers to consumers. Certain schemes may be found in the Community and other countries.

Title: Verolme Liquefied Natural Gas Carrier	Project No.: 12.04/77
Contractor: Naval Project Development Company (Verolme) Address: Blaak 101 NL - Rotterdam Technical director (or person to contact for further information): Dr A K Winkler	Telephone No.: (010) 112 670 Telex: 26054

Aim of the Project:
Development of the Verolme LNG carrier.

Project Description:
The project envisages the development of a very large gas carrier over 3000 000 m^3. The design will consider already proven concepts.
- The hull of the ship is of conventional container shal design, adapted to the need of the cryogenic containment system and to the requirement of shallow draft.
- the containment system consists of several vertical aluminium cylinders.

The project is divided into 2 main phases:
Part I: Ship
- remaining basic development of ship hull
- model testing a ship hull

Part II: Cryogenic system
- design of cryogenic containment system
- 3D structural analysis
- design of ancillary cryogenic equipment
- design of operating equipment
- model testing and construction of model and related equipment.

State of Project:
The project was stopped in December 1979.

Results and applications:
Part I: Ship
Softwear design work has been completed to provide the basic data for any future detailed design work to be done by a yard. A model (1:42 scale) was tested and extrapolation made to determine the required horse power. Manoeuvrability was tested giving good results.
Part II: Cryogenic system
The results of this development are listed below:
- general arrangement
- cylindrical container lay out
- container details (pipelines, stairways, platform, manhole, ASO)
- supporting grid
- load and discharge diagram
- piping general arrangement.

Lloyd's has accepted the lay out of the cryogenic system underneath the desk. Furthermore, "large-scale structural analysis" was performed by Lloyd's and the behaviour of the integrated containment system was found to be satisfactory.

Title: Insulation and barrier system for marine transport and storage of LNG	Project No.: 12.05/78
Contractor: Shell Research Ltd Address: 30, Carel van Bylandtlaan NL - 2501 AN Den Haag Technical director (or person to contact for further information: H L Beckers	Telephone No.: (070) 779 111 Telex: 31005

Aim of the project:

To develop a rational insulation system design suitable for containing LNG in a ship or storage tank.

Project Description:

To develop a system based on sprayed polyurethane foam applied to the inside shell of a storage tank. To prove that the system functions as required by practical tests in a laboratory and in test tanks.

State of Project:

Project abandoned in June 1981 after failure of test tank.

Contract completed in March 1982 with completion of a technical audit and submission of technical audit and submission of technical reports.

Results and applications:

No further development undertaken or application made of this work.

ENERGY SOURCE

Title: High density underwater energy source	Project No.: 13.05/78
Contractor: COMEX Industries Address: 36, Boulevard des Océans F - 13009 Marseille Technical director (or person to contact for further information): Y. Durand	Telephone No.: (91) 41 01 70 Telex: 401755

Aim of the Project:

To develop and test an air-independent power system for underwater operations based on the Stirling Engine technology.

Project Description:

The Stirling Engine is an external combustion engine burning fuel and pure oxygen in a pressurized combustion chamber supplying heat to a closed circuit working gas acting on a four-cylinder double action piston engine. The project consisted of building and testing a prototype power system capable of producing up to 20 Kw electricity to a depth of 200 m.

State of Project:

The project has been completed in December 1982, after extensive endurance tests, thus demonstrating the feasibility of an alternative underwater energy source.

Results and applications:

The non-availability of powerful and long-endurance civilian energy sources has limited the development of surface-independent solutions for underwater working system. The results have allowed us to proceed with a real-size demonstration of a large manned, diver lock-out submarine, the "Saga I". This 300 tonne prototype, to be completed by August 1986, includes two V4-275R-80Kw Stirling Engines and a liquid oxygen storage system, capable of surface-independent operations for durations of up to 25 days, in actual conditions and down to 600 m, thus preparing for the next generation of underwater developments.

STORAGE

Title: Underwater oil storage tank	Project No.: 15/75
Contractor: Tecnomare SpA Address: S. Marco 2091 I - 30124 Venezia Technical director (or person to contact for further information): Mr Rodighiero	Telephone No.: 70822 Telex: 41484

Aim of the Project:

Main targets of the underwater oil storage research project are described as follows:

- to obtain and/or transfer the basic methods and techniques to the design of large offshore structures made of steel or reinforced concrete
- to develop the design of a 300m water depth underwater oil storage tank.

Project Description:

The design of the underwater oil storage tank derives from the integrated solution of different problems regarding:
storage process system, structure configuration, design methods, installation procedures, construction methods, operative life and maintenance procedures.

Different design aspects have been analysed in detail so that computer procedures, theoretical studies, model tests and field tests have been carried out.

State of Project:

Complete.

Results and applications:

The results of the experimental tests performed confirm the feasibility of the concept with special reference to the fabrication, insulation and operating procedures.

The system of the underwater storage oil tank has a configuration flexible enough to meet a wide range of storage capacities, installation water depths, meteoceanographic conditions and types of connection production and mooring systems.

It represents a valid solution for exploitation of marginal fields over 100m and for large fields in deep water where floating storages become uneconomical.

Title: Research on the use of oil reservoirs in fractured rocks for the storage of liquid and gaseous hydrocarbons	Project No.:16/75
Contractor: Agip SpA Address: P.O. Box 12069 I - 20120 MILANO Technical director (or person to contact for further information): Prof. G.L. Chierici	Telephone No.: (2) 520 40 86 Telex: 310246 ENI I

Aim of the Project:

To find out whether oil recovery from the fractured, heavy oil reservoir of Gela, Italy, can be enhanced by replacing the natural water drive mechanism with the injection of either natural gas or carbon dioxide.

Project Description:

Laboratory tests have been performed on cores to assess how gas drainage improves oil recovery in comparison with water imbibition, and how injected gas affects the thermodynamic characteristics of reservoir oil. A three-dimensional, black oil numerical model has been employed to evaluate the improvement in oil recovery and oil production rate that can be achieved under different gas injection schemes. The model had been previously validated by matching field past history; the future field performance by natural water drive has also been evaluated, to provide a basis for comparison.

State of Project:

Completed.

Results and applications:

From the results of the numerical model study it appears that oil recovery from Gela field can be significantly improved by substituting the natural bottom water drive with crestal injection of either natural gas or carbon dioxide. As the results of a field pilot in a large field such as Gela are difficult to interprete and evaluate, the pilot testing of high pressure gas injection has been switched to the nearby Ponte Dirillo oil field, whose characteristics are very similar to those of Gela (Project No. 05.14/79)

Title: Underground cryogenic storage in rock - Realisation of a test gallery	Project No.:17/75
Contractor: GEOSTOCK Address: Tour Aurore, CEDEX 5 F - 92080 PARIS DEFENSE Technical director (or person to contact for further information): M. Alain BOULANGER	Telephone No.: (1) 778 53 53 Telex: 610898

Aim of the Project:

To demonstrate the feasibility of an underground cryogenic storage in limestone for liquefied gas.

Project Description:

The project is divided in several phases:

1. Laboratory test in limestone to appreciate the feasibility of the test plant;
2. To design, build and operate a test gallery in limestone. The 5 000 m3 test gallery is to be built in limestone in Lavera near Marseilles;
3. To interprete the results of the test gallery and conclude on the feasibility of an industrial plant in limestone.

State of Project:

For economical reasons connected with the industrial development of this technique, the project was stopped in 1977 once the first phase was completed.

Results and applications:

The test laboratory allowed to conclude that it is theoretically possible to store LPG and LNG in limestone, but a test plant is needed to confirm this conclusion.

Title: LNG storage in salt cavities	Project No.: 14.02/77
Contractor: KBB-Ruhrgas Address: Postfach 28 Muttopstrasse 60 D - 4300 Essen 1 Technical director (or person to contact for further information): Mr Zündel/Mr Lindemann	Telephone No.: (0201) 1841 Telex: 01- 57818

Aim of the Project:

Investigation of LNG storage in salt caverns.

Project Description:

In 1974 the possibility in principle of storing low temperature liquids has been experimentally proved in a test cavern of about 1 m3.

In the framework of the project, the thermodynamic working properties of LNG in the cavern and the unstable 3-dimensional temperature field in the neighbourhood of the cavern has been established. Conformity with experimental results was satisfactory. Between calculated and observed rock displacement, a qualitative conformity has been obtained.

The forming of cracks orthogonal to the cavern suface in salt rock has been clarified by extrapolating the known geophysical data of the salt. It has been shown that because of the insufficient knowledge of the material properties of the rock salt at low temperature and the complex mathematical boundary and starting conditions utilising the available computation method, no precise predicitons regarding thermodynamic and rock mechanical behaviour in large caverns can be without intensive experimental research.

The unstable temperature in the neighbourhood of the cavern for the storage of LNG has been calculated with a model cavern of 5 000 m^3. The basis was a quasi-stationary storage at high pressure without discharge of boil-off gas. Some larger cracks in the cavern wall which developed during the tests did not influence the operational ability of the cavern. In practice it is important that crack-forming should stop at a certain time and distance from the cavern wall.

Because of the high material streams involved in the rapid loading of LNG tankers, it is necessary for LNG to be loaded in the same time using a greater number of tubes. A storage system with seven caverns has been developed, flow diagrams have been prepared and working procedures developed. A detailed economic investigation has been commenced to yield a qualitative indication of the comparison between salt cavern, LNG storage and a conventional one with storage tanks. Salt caverns at the coast and distant form the coast have been taken into consideration.

Title: Development and optimisation of hydrocarbon storage facilities	Project No.: 14.03/78
Contractor: Sir Robert McAlpine & Sons Ltd. Address: 40, Bernard Str. UK - London WC1N 1LG Technical director (or person to contact for further information): Dr T L Shaw	Telephone No.: 01 837 3377 Telex: 22308 G

Aim of the Project:

To consider world experience with all forms of hydrocarbon storage facilities, with a view to assessing how the particular merits of concrete could best be applied for the safe containment of liquids at atmospheric, and at low and cryogenic temperatures.

Project Description:

A detailed literature appraisal was first made and contact established with tank owning and operating companies. Reported incidents were assessed and considered in terms of their significance for a variety of tank designs in prestressed concrete using present day methods, materials and safety criteria.

State of Project:

Phase 1 completed in April 1982. Phase 2 postponed pending renewed industrial interest in cryogenic storage of liquefied hydrocarbons.

Results and applications:

It was concluded that there was a good structural argument for the safe containment of cryogenic liquids using a single-wall prestressed concrete tank principle. It is argued that liquid containment and security against external forces should be done by separate structures, hence a twin-wall installation is necessary. The proposed "Phase 2" of project is a design in which the inner wall has nominal strength, compared with that which surrounds it.

Title: SISSAC - self-installing subsea storage and articulated column	Project No.: 14.04/78
Contractor: EMH and Halcrow Ewbank Address: 196, Boulevard de la Colline F - 92213 St Cloud Cédex Shortlands UK - London W6 8BT Technical director (or person to contact for further information): R. Aftalion	Telephone No.: 771 91 22 Telex: 204586

Aim of the Project:

Development of a new system for subsea storage of oil.

Project Description:

The system combines an underwater concrete storage tank with the widely used articulated loading column. The system is adapted to water depths from 120 to 220 m and to storage capacities from 400 000 bls to 750 000 bls and installation in such locations as the North Sea.

- The storage tank is built in concrete and the loading column in steel.
- The SISSAC provides the column transportation to the site using the concrete tank as a barge.
- The SISSAC is installed by filling tank cells and column compartments in a given sequence.

State of Project:

The project was completed in August 1982.

Results and applications:

The project enabled the technical and economic validity of the concept to be proved. All phases of SISSAC life have been considered in the project: construction, launching, tow-out, installation and 20 years' exploitation. Particular attention has been given to the most critical and unclassical aspects of the project, i.e. the installation procedure, the foundation design and equipment in the tank and in the column.

The system proved its economic interest for remote oil fields, where neither pipelines nor big storage structures are viable.

Title: Unlined concrete storage facilities for liquefied natural gas	Project No.: 14.06/78
Contractor: Taylor Woodrow Construction Ltd. Address: 345 Ruislip Rd, Southall UK - Middlesex UB1 2QX Technical director (or person to contact for further information): P.B. Bamforth (Project Co-Ordinator)	Telephone No.: 01 575 4578 Telex: 24428

Aim of the Project:

To provide the technical data required to support the design of unlined, as well as lined, prestressed concrete facilities for the storage of liquefied natural gas (LNG).

Project Description:

Tests have been carried out to identify the concreting materials and mix proportion most suited to cryogenic applications. Selected concretes were comprehensively tested at temperatures down to -165°C to determine the engineering properties required for design. A range of reinforcing steels and prestressing systems were also investigated. Finally, prestressed concrete elements were tested under conditions of live and thermal loading.

State of Project:

The test work has been completed and a final report prepared.

Results and applications:

Results are available which show the influence of cryogenic temperatures on a range of concrete properties including: permeability; compressive strength; tensile strength and strain capacity; flexural strength and strain capacity; elastic modulus; creep; bond to steel; thermal contraction; specific heat; resistance to thermal cycling.

Data on properties of steel for rebar and prestressing has also been obtained including: elastic modulus; yield stress; ultimate tensile stress; strain at failure; impact transition temperature (Charpy and full bar tests).

Light-weight concrete was identified as being particularly suitable for cryogenic applications and the performance of light-weight prestressed beams has been recorded.

<table>
<tr><td><u>Title</u> :

Permanently moored tanker used for deep water storage and off-loading</td><td>Project No. : 14.07/79</td></tr>
<tr><td><u>Contractor</u> :
SBM (UK) Limited

<u>Address</u> :
Northumberland House, 2 a King Street
Twickenham
UK - Middlesex TW1 3SN

<u>Technical director (or person to contact for further information)</u> :
Mr. Bray/ Mr. Dyer</td><td><u>Telephone No.</u>:
(01) 891 3434

<u>Telex</u> :
28306</td></tr>
</table>

Aim of the Project :
The purpose of this study was to examine the feasibility of a mooring system suitable for a converted tanker in a water depth of 400 m under severe environmental conditions typical of the North Atlantic.

Project Description :
In order to widen the possible future applications, the effects of a milder environment typical of the Mediterranean have also been considered. This project had three key objectives :

- to develop a permanent tanker mooring system for use in deep water and rough sea conditions
- to examine the deep water system in milder sea conditions
- to develop the system to a point where subsequent commercial development can begin.

State of advancement :
Completed.

Results :
a) Single Buoy Moorings is confident that engineering of the five mooring concepts developed could take place by scaling-up proven components with no novel technology required.
b) The required hard ware development for the North Atlantic environment would be least for DSM II (an articulated anchor leg with catenary legs) and the Bow turret because of their lower mooring forces.
c) For Mediterranean conditions conventional systems could be designed and engineered without any significant scaling problems. It concerns mainly :
   - SBS system (a buoy anchored by six or eight chain catenaries)
   - SALS system (an articulated anchor leg connected to a base on the seabed)
d) The vessel size (within a practicable range) is not the main parameter for mooring system forces in 400 m water depth.
e) The slow drift behaviour of permanently moored tankers is over-estimated when extrapolating previous tests with smaller waves. However the other motions and the general level of loads have been well predicted using techniques developed by SBM.
f) Installation techniques were studied in sufficient depth to show that all the systems were feasible to install in deep water.

Title: Underground cryogenic storage for LPG and LNG in rock	Project No.: 14.10/80
Contractor: Distrigaz ave des Arts 31, B - 1040 Brussels Technical Director: Paul SOILLE & Geostock Tour Aurore 5, Cedex 5 F - 92080 Paris Technical Director: Alain BOULANGER	Telephone No.: (2) 230 50 20 Telex: 23317 Telepnone: (1) 778 53 53 Telex: 610898

Aim of the Project:

To demonstrate the feasibility of an underground cryogenic storage in clay for liquefied gas.

Project Description:

The project is divided in several phases:

1. Laboratory test in clay to appreciate the technical feasibility of the test plant;

2. To design, build and operate a test gallery in Boom clay. The 200 m3 test gallery is to be built in Boom clay at Schelle near Antwerpen (Belgium);

3. To interprete the results of the test gallery and conclude on the feasibility of an industrial plant in clay.

State of Project:

The project is completed.

Results and applications:

The feasibility of such a storage is demonstrated. This technique is applicable for the storage of LPG (-45°C) in any kind of rock and for LNG (-162°C) for Boom clay type formations.

Title: Construction techniques in limestone for cryogenic storage	Project No.: 14.13/82
Contractor: Cavern Systems Dublin Limited Address: 16, Upper Pembroke Str. IR - Dublin 2 Technical director (or person to contact for further information): F.C.M. Barter	Telephone No.: 789 400 Telex: 33203

Aim of the Project:

To extend and develop existing underground rock cavern technology with a view to constructing 100,000 tonnes of underground refrigerated LPG storage in Dublin Bay Limestone.

Project Description:

The project includes the following main elements:

- Detailed geological investigation and geological appraisal of the port area of Dublin
- Comprehensive feasibility studies of the environmental, hazard, technical design, and economic aspects
- Construction of an exploratory shaft to an approximate depth of 110 m. and performing comprehensive tests
- Construction of scaled down test cavern and performing comprehensive tests to determine whether full scale cavern proceeds and the final design configuration.

State of Project:

The geological/geotechnical appraisal and feasibility study stage of the project has been successfully completed. Commencement of the construction of the exploratory shaft and scaled down cavern has been delayed pending the receipt of planning permission. Although the relevant statutory planning authority has granted planning permission, the outcome of a planning appeal hearing is currently awaited.

Results and applications:

The geological, geotechnical and geophysical appraisal established the overall suitability of Dublin Bay Limestone for refrigerated LPG storage. The feasibility studies established the overall environmental acceptability, the preliminary design configuration and geometry and the likely economics of the project. These results will be applied to the remaining construction stage of the project leading towards the eventual construction of 100,000 t. of refrigerated LPG storage in Dublin and similar developments in other EEC countries.

Title: Permanent mooring of a floating unit in deep water by means of a multi-articulated column	Project No.: 14.14/82
Contractor: GERTH/EMH Address: (EMH) 196 Bureaux de la Colline F - 92213 St Cloud Cédex Technical director (or person to contact for further information): Radu Aftalion	Telephone No.: (1) 771 91 22 Telex: 204586

Aim of the Project:
Development of the articulated column technique for permanently mooring a floating unit in depths of water of from 600 to 1000 metres and environments of the North Sea or Mediterranean type.

Project Description:

General analysis:
- definition of the requirements of the oil companies
- a review of the different mooring systems developed or now undergoing development for great depths of water
- definition of the representative conditions of environment of the two types of site (waves, wind, currents, soil)
- definition of typical production schemes for study of fluid transfer systems
- review of the fluid transfer systems available (low and high pressure)

Preliminary projects:
- definition of the general structure of deep water columns
- development of methods for predimensioning and analysing the behaviour of the structures

Specific studies:
- application of weight-reducing materials for the construction of high pressure buoys
- analysis of fluid transfer systems along the column

State of the Project:
Feasibility studies and model tests will be performed in 1984 and 1985.

Results and applications:

The general analysis phase has revealed the advantages of permanent moo mooring systems on a multi-articulated column for the development of deep water offshore fields. The reason is that there are many applications, ranging from simple buffer storage to the accumulation of the support functions of equipment for controlling and processing production, storage of the oil produced and its discharge via shuttle tankers moored in pairs or in tandem.

The main alternatives defined both for the architecture of the mooring structures and the fluid transfer system have demonstrated the existence of a wide variety of solutions for the technical problems set that will be selected to suit each specific case studied in the rest of the project.

Title : DEVELOPMENT OF A NEW STORAGE TECHNOLOGY FOR LPG	Project n° : 14.15/82
Contractor : GERTH/TECHNIGAZ Address : 4, av. de Bois Préau 92500 RUEIL-MALMAISON FRANCE Technical director (or person to contact for further information) : Mr Gilbert BLU or Mr Bernard MERCIER	Telephone n° : (1) 749.02.14 ext. 2288 or 2747 Telex : 203 050 F

AIM OF THE PROJECT

The purpose of this project is to develop a new technology for shore storage of liquefied gases at atmospheric pressure and of high capacity (100,000 m3), applying the "membrane" technique, separating the support function from the sealing function ensured by membrane laid on the insulation. The objective of this project is to study the application on LPG tanks of a composite membrane that is much less expensive than the metal membranes used with LNG.

PROJECT DESCRIPTION

The project consists in designing, building and testing a tank with a capacity of about 2,000 m3, this being a scale representative of subsequent industrial structures, built in accordance with the membrane technique known as GMS (gas membrane system). This pilot tank comprises the following elements :

- a prestressed concrete shell (slab + cylinder) with a wall thickness of 400 mm. The dome of the tank consists of a metal envelope coated with a layer of concrete
- a steam-barrier coating on the internal face of the tank
- an insulating wall made up by juxtaposing 100 mm thick panels of PVC or polyurethane foam
- a "Triplex" sealing membrane consisting of a sheet of aluminium 70 micrometres thick placed between 2 layers of glass fabric. This envelope, which is both perfectly gas-tight and liquid-tight, is applied to the panels in the works. After laying the panels inside the tank, the continuity of the Triplex membrane is ensured by joints in the same material.

The improvements expected from this technique are :

- insulation without LPG gas content : the insulation space is in fact filled with inert gas which is continuously analysed
- very easy access to work on and repair the tank.

STATE OF PROJECT

The following phases of the work have been completed :

- following laboratory tests, determination of the specifications and assembly procedures
- homologation of the materials
- construction and acceptance testing of the prestressed concrete tank within the Flandres CFR refinery at Dunkirk
- ordering of the compressors and pumps
- the insulation and equipment will be installed during 1984, and the tests will start at the end of the year.

MISCELLANEOUS

Title: Development of a total structural monitoring system for offshore platforms	Project No.: 15.03/77
Contractor: I.I.R.S. Address: Ballymun Rd, IR - Dublin 9 Technical director (or person to contact for further information): Mr Phelim P. Rooney	Telephone No.: (01) 370101 Telex: 25449

Aim of the Project:

To develop a total structural monitoring system for offshore platforms.

Project Description:

Accelerometers, a wave staff and a computer based data acquisition system were installed on Kinsale Alpha and the records were analysed to determine if the structural integrity of the platform could be monitored. A design safety factor assessment was also considered.

State of Project:

Completed 1981.

Results and applications:

The results indicate that such a system cannot replace current underwater inspection and maintenance practice. No further applications of the system were made by IIRS. The main problem areas were:

- a lot of instrumentation is necessary (mostly underwater)
- the system is sensitive to deck load, marine growth, etc.
- maintenance and regular calibration of the system would be expensive. The project did not provide valuable feedback to IIRS on safety levels inherent in the design of the Kinsale jackets.

Title: Diagnostic methods for offshore structures	Project No.: 15.06/78
Contractor: Tecnomare SpA Address: S. Marco 2091, I - 30124 Venezia Technical director (or person to contact for further information): Mr V Banzoli	Telephone No.: 708622 Telex: 410484 MAREVE I

Aim of the Project:

To establish the state of integrity of offshore structure by means of a portable instrumentation system; to specify a permanent instrumentation system and to develop the relevant software; to analyse new design methodologies and their mutual influence with monitoring inspection activities.

Project Description:

The portable instrumentation system is based on the underwater monitoring of vibration induced by local excitation. The system is composed of three underwater accelerometer heads and electronics for data acquisition, an underwater electro-hydraulic exciter and a surface container equipped with data recording and control electronics. Besides the design and construction of the portable monitoring system during the project development, other activities were performed:

- issue of a set of computer procedures and programmes to update the fatigue accounting and to simulate the propagation of cracks, by using strain measurements in a few points of the structure;

- examination of different design methodologies based on "damage tolerance approach" in order to establish their applicability to the design of offshore structures.

State of Project:

Completed.

Results and applications:

The portable instrumentation system has been tested at sea, while the computer procedure for fatigue accounting and crack growth speed prediction have been tested on an actual platform.

Title: Meteoceanographical and structural data acquisition to improve platform design	Project No.: 15.07/78
Contractor: Agip SpA Address: Agip SpA, Offshore Dept. 20097 S. Donato Milanese, I - Milan Technical director (or person to contact for further information): Mr M Magni	Telephone No.: 02/52 027 220 Telex: 310246 ENI I

Aim of the Project:

- To verify scatterings of the results of present calculations from actual measured values;
- To calculate the fatigue damages and to make previsions on the remain fatigue life;
- To individuate irreversible failures of the structures.

Project Description:

The system is programmed to perform and record environmental conditions, structural stresses and measurements of the dynamic response of the platform.

State of Project:

Open.

Results and applications:

To date all the equipment has been tuned and is working successfully. Some interesting results have been gained about the dynamic response of the platform and its account fatigue life. The acquisition period is expected to terminate towards the end of 1986. At that time we feel the other aspects of the study will be completed.

Title: Development of measurement system to optimize computer programmes and to monitor dynamic behaviour of offshore platforms	Project No.: 15.08/78
Contractor: SYMINEX Address: 2, Boulevard de l'Océan F - 13275 Marseille Cédex 9 Technical director (or person to contact for further information): A.J. Kermabon	Telephone No.: (91) 73 90 03 Telex: 400563

Aim of the Project:

The aim of the project has been to set up a procedure and equipment to measure the dynamic characteristics of offshore structures with two objectives in view:

- to optimize theoretical models of the structure and to associate realistic damping values
- to detect possible cracks on these bases.

Project Description:

In the first stage, computation programmes were chosen and numeric results obtained. A model size platform was built, on which verification of a theoretical model was provided (before and after the model was damaged). Several types of excitation systems were studied and finally an operational equipment was developed (local and global exciter, accelerometric box, acquisition unit). During the second stage, offshore tests were carried out in real conditions in order to set up the equipment, to define a measuring and interpretation procedure and to find out the limits of the crack detection.

State of Project:

The project is now quite complete; all the equipment has been developed and is now operational. Further research projects should be carried out, in order to improve them.

Results and applications:

At the present stage, the project has resulted in two specific techniques:

- model analysis: this technique did not really give the answer to the initial hopes
- vibrodetection: the purpose of which is to detect and localise significant cracks on an offshore jacket. Many offshore tests - as in the Arabian Gulf in 1982 - proved the reliability of this technique.

Title: Meteoceanographical - structural measurement system for the safety of the Nilde single anchor leg storage Sicily Channel	Project No.: 15.11/80
Contractor: Agip SpA, Offshore Dept. Address: 20097 S. Donato Milanese I - Milano Technical director (or person to contact for further information): Mr M Magni	Telephone No.: 02/52 027 220 Telex: 310246 ENI I

Aim of the Project:

- To verify scatterings of the results of present calculations for the SALS system from the actual measured values;

- To compare the real measured values with the model results.

Project Description:

The project has been conceived in order to record and process meteoceanographical, structural and dynamic data collected by appropriate sensors and a data acquisition system.

State of Project:

Open.

Results and applications:

To date, the data acquisition period has been completed. The final onshore data processing, together with the conclusive dynamic study, is about to start. The conclusion of the project is expected towards the end of 1984.

Title: Development of methods of rehabilitating damaged offshore concrete structures	Project No.: 15.16/80
Contractor: Taylor Woodrow Construction Ld. Address: 345 Ruislip Road, Southall UK - Middlesex Technical director (or person to contact for further information): Dr A McLeish	Telephone No.: 01-575 4856 Telex: 24428

Aim of the Project:

The aim of the research project was to develop practical repair procedures for damaged offshore concrete platforms. Damage including broken reinforcement and prestressing tendons in addition to cracked or crushed concrete was considered.

Project Description:

The research project was carried out under seven main headings:

- Assessment of the extent of damage
- Study of methods of access to the damaged area
- Preparation of the damaged area
- Reinforcement repairs
- Repairs to prestressing
- Replacement of concrete
- Large scale trials

State of Project:

Complete.

Results and applications:

Repair procedures and materials have been assessed in large scale repair trials. The methods developed are practical to carry out in the offshore environment and have been shown to perform well.

Title: High energy welding	Project No.: 15.22/81
Contractor: British Underwater Engineering Ltd Address: 12/18 Grosvenor Gardens UK - London SW1W 0DW Technical director (or person to contact for further information): J M Lowes	Telephone No.: (0229) 250 80 Telex: 65147

Aim of the Project:

To extend the use of existing high energy welding technology to a depth of 1 000 ft. It is also intended to develop the necessary support systems and to investigate the design of a sleeve with a double high energy capacity. This would be applicable to an all welded connection using a minimum of handling equipment.

Project Description:

The development of the project will be submitted into 3 major work areas:

1. Welding technology

   To establish the parameters for 2 pipe specifications and demonstrate by simulation testing the acceptability of the process at a depth of 1 000 ft sea water. A demonstration would be arranged with certifying authorities for a full-scale test to be performed remotely at a depth of 1 000 ft sea water.

2. Systems

   The following aspects will be studied:
   - pipe sizing tool
   - multi-size machining capability
   - ultrasonic inspection system
   - bag deployment monitoring
   - corrosion inhibition
   - remote detonation unit
   - umbilicals and flush and dry sequencing unit.

3. Feasibility of a double welded sleeve

   The following particular aspects will be investigated:
   - overall sleeve design to accomodate the various operational requirements
   - seal design and arrangement
   - charge design
   - corrosion problem
   - handling problems

Results and applications:

No significant progress due to lack of participants from oil industry for financial support.

Title: Optimization of the construction and inspection procedures for storage and transportation of hydrocarbons, 1983-1984	Project No.: 15.23/81
Contractor: ATB (Acciaieria e Tubificio di Brescia) Address: Box 308 I - 25100 Brescia Technical director (or person to contact for further information): A Berzolla	Telephone No.: (030) 53361 53461 Telex: 301636 340269

Aim of the Project:

Verify the new welding technologies with submerged arc welding applied on 9% nickel steel, employed for the containers of storage, transportation and processing of hydrocarbons, in the thickness range 18-48mm.

Project Description:

On the basis of chemical and mechanical properties, wires and fluxes have been subdivided into three weld metal groups:

a) for weld metal type 70 Ni - 20 Cr - 2.5 Nb
b) for weld metal type 60 Ni - 20 Cr - 9 Mo - 3.5 Nb
c) for weld metal "modified stainless steel" type (for instance: 18 Cr - 13 Ni - 7 Mn 3 W)

State of Project:

Completed.

Results and applications:

On the basis of the various tests performed, our evaluation of the results obtained is the following:

- to weld steel plates type 9% nickel with submerged arc process, the best mechanical characteristics can be obtained using materials belonging to Group A (type 60 Ni - 20 Cr - 9 Mo - 3.5 Nb). With this type of weld metal we obtain (in 80% of cases) the minimum characteristics required for the base metal
- the best combination, from the point of view of both operational and mechanical characteristics, is the wire-flux combination E9-F2 of Group B
- another satisfactory combination is E11-F4 of Group C
- considering the economical aspect of this matter, we would like to underline hereunder the price difference in the three types of deposits, namely:

  Group A: LIT 40 000/50 000 (for each kg of weld deposit)
  Group B: LIT 60 000/80 000 (for each kg of weld deposit)
  Group C: LIT 20 000/25 000 (for each kg of weld deposit)

Title: Development of a Reliability Analysis System for Offshore Structures (RASOS)	Project No.: 15.24/81
Contractor: I.I.R.S Address: Ballymun Rd., IR - Dublin 9 Technical director (or person to contact for further information): Mr Phelim P. Rooney	Telephone No.: (01) 370101 Telex: 25449

Aim of the Project:

To develop a reliability analysis system for offshore structures

Project Description:

The project involves:

- Characterisation of stochastic variables for Irish waters
- Development of reliability system (certification procedures and computer programme)
- Development of quality assurance and planned maintenance programmes
- Application of RASOS to an offshore installation.

State of Project:

Current.

Results and applications:

Work is ongoing and data gathered since 1981 is to be presented in a readily useable format. The Rules and Procedures developed to date shall be applied to an Irish offshore installation and the overall results reported upon completion.

Title: Development of a long range self-contained submarine power supply	Project No.: 15.30/82
Contractor: Sté Bertin & Cie Address: BP 3, F - 78373 Plaisir Cédex Technical director (or person to contact for further information): Mr S Galant	Telephone No.: (3) 056 25 00 Telex: 696231

Aim of the Project:

To design and to test a long range self-contained submarine power supply with peak electric power capabilities above 100 KWhrel and capacity well above 1 000 KWhrel

Project Description:

The project is divided into 4 phases:

Phase 1: Feasibility study of a liquid methanol-pressurized oxygen combustion chamber, coupled with an Organic Rankine Cycle for electricity generation and a heat exchanger configuration leading to full liquefaction of the combustion gases.

Phase 2: Laboratory tests of the liquefaction system.

Phase 3: Full scale tests of combustion chamber design

Phase 4: Full scale tests of the self-contained submarine power supply both under onshore and offshore conditions.

State of Project:

Phases 1 and 2 have been successfully completed at the end of 1983. Programme proposals have been submitted to adequate French and EEC organisations to financially support Phase 3.

Results and applications:

Phases 1 and 2 results have shown that:
- the combustion chamber should be designed by making an extensive use of aeronautic methodology: the minimum thermal output should stand around 150 to 200 KWthermal;
- it is necessary to recirculate exhaust gases in order to maintain gas temperature within acceptable ranges (viz. below 2000 K);
- an organic Rankine cycle machine will yield a 20% overall thermo-mechanical conversion efficiency;
- full liquefaction of combustion gases is possible with conventional heat exchanger configurations working at 60 bars, 15°C average liquefaction conditions.

Title: Cast steel nodes for fixed offshore structures	Project No.: 15.32/82
Contractor: Britoil PLC Address: 150 St Vincent Str. UK - Glasgow G2 5LJ Technical director (or person to contact for further information): Mr T.E. Evans, Study Manager	Telephone No.: 041 204 2566 Ext.5632 Telex: 777633

Aim of the Project:

To carry out an independent evaluation of the properties of prototype cast steel node for offshore structures and to define defect acceptance levels and mechanical properties for specification purposes.

Project Description:

The project encompasses the development work needed to enable cast steel nodes to be specified with confidence for use in offshore steel structures. Phase 1 consisted of design and production of a full-size prototype launch brace node as a casting. Phase 2 comprises 100% surface and volumetric non-destructive testing to determine casting, quality, welding and weld repair demonstration and evaluation and mechanical testing of a large number of representative samples taken from different parts of the casting. A limited programme of fatigue testing is planned for Phase 3.

State of Project:

Phases 1 and 2 have been completed and a project report covering the work performed in Phases 1 and 2 is in preparation.

Results and applications:

Phase 1 has demonstrated that a large complex node can be produced as a sound casting to a high standard and to a realistic production schedule.

Phase 2 has demonstrated that the surface condition of the casting was good and that ultrasonic testing is in general a satisfactory technique for detecting and locating buried defects such as shrinkage porosity and hot tears. The stub-to-tubular weld and weld repair trials have been successfully completed. Mechanical property values in a large casting have been shown to vary within wide limits and the data obtained can be utilised for the definition of minimum property values for specification purposes.

Title: The mechanical echo method of diagnostic for offshore structures	Project No.: 15.33/82
Contractor: Tecnomare SpA Address: S. Marco, 2091 I - 30124 Venezia Technical director (or person to contact for further information): V. Banzoli	Telephone No.: 708 622 Telex: 410484

Aim of the Project:

To achieve a better capability to diagnose the actual safety conditions of steel structures, by means of a new method based on the elastic wave propagation and reflection.

Project Description:

The basic principle of the proposed method is that an elastic wave propagating along a structure shares its energy between a transmitted wave and a reflecting one when it reaches a discontinuity in the structure (e.g. a crack). Measuring the time of return back and the energy of the reflected wave, it should be possible to locate and quantify the discontinuity.

The project is subdivided into the following main subjects:

- acquisition of basic theoretical knowledge
- theoretical and laboratory experimental study of the method
- design and procuring of a subsea portable monitoring system
- field test of the portable system on a real platform.

State of Project:

The theoretical study of the method and the laboratory tests are completed.

Results and applications:

At present the main results achieved consist of the issue of the computer programmes for simulating the response of a cylindrical structural element and of the definition of the techniques for the analysis and treatment of the measured signals (including the issue of the relevant computer programmes).

Title: Subsea oil loading system	Project No.: 15.34/82
Contractor: AEG-Telefunken Address: Steinhöft 9, D - 2000 Hamburg 11 Technical director (or person to contact for further information): Dr Wilke	Telephone No.: 040/3616-1 Telex: 211868

Aim of the Project:

Study of a concept of a subsea oil loading system and in particular for engineering of the individual components. The aim of the R & D project is functional and manoeuvring trials of the system.

Project Description:

The system can be described as follows: the tanker is manoeuvred into position and has a moonpool close to its manifold. This moonpool enables a re-entry unit at end of loading hose which is lowered into the water and guided to the Pipeline End Manifold and couples with it to make the hose link for oil loading. Coupling is effected by remote control. PLEM is connected via a pipeline with production platform.

State of Project:

Project in progress. Development, construction and testing as a full scale model in a harbour basin in 1984.

Results and applications:

The system is expected to give many advantages: low expense in comparison with the highly complex offshore loading systems required today; favourable handling times; high availability even under rough weather conditions; economic operation at great water depths and in marginal oil fields; maintenance and service not dependent on weather; elimination of risk of collision with loading facility; reduced storage capability.

Moreover, SOLS can be used for the development of production areas not accessible to surface loading systems and the addition of the dynamic positioning system gives the tanker major manoeuvrability benefits. SOLS can economically pump a tensid fluid into a subsea well and carry back crude oil from another well some miles away. Application of heavy ballast transportation appears realistic.

Title: A field investigation into the soil-structure interaction of a foundation system during the early life of an offshore oil production platform	Project No.: 15.35/82
Contractor: The British Petroleum Co. PLC Address: Britannic House Moor Lane UK - London EC2Y 9BU Technical director (or person to contact for further information): W.J. Rigden	Telephone No.: (01) 920 8264 Telex: 888811

Aim of the Project:

The aim of the project is to obtain early data on the behaviour of the foundation of an offshore oil production platform.

Project Description:

BP Petroleum Development Ltd have installed a production platform in 186m water at Magnus Field in block 211/12 of the UK sector, North Sea.

A separate project, 06.14/81, provides for the investigation of foundation behaviour. This project allows the early installation of a temporary data logger to provide monitoring of foundations at the critical period immediately following the positioning of the platform tower.

State of Project:

The temporary data logger was used to record information in April 1982, following the positioning of the platform tower. The temporary data logger was disconnected in July 1983 in readiness for connection of the permanent facility.

Results and applications:

Results and analysis of the data is continuing and shall be reported as part of the project 06.14/81.

Title: Development of techniques for the inspection of offshore concrete platforms	Project No: 15.37/82
Contractor: Taylor Woodrow Construction Limited Address: 345 Ruislip Road Southall UK - Middlesex UB1 2QX Technical director (or person to contact for further information): Dr R D Browne	Telephone: (01) 578 2366 Telex: 24428

Aim of the Project:

To develop improved techniques for the inspection of concrete structures above and below sea level. These would include measurement tools and assessment techniques for rapidly quantifying the corrosion state of reinforcement and inspection methods to aid the assessment of pre-stressing steel corrosion.

Project Description:

The programme will examine the development of systems based on electro-chemical measurement, ultrasonic techniques, development of current density monitors, chloride analysis and measurements of reinforcement cover to enable the corrosion rate of reinforcement to be assessed. The second part of the project, which will look at assessment techniques for prestressing steels, will investigate developments based on acoustic omission for corrosion assessment and NDT techniques for leak and void detection.

State of Project:

Not commenced

Results and Applications:

-

Title : ELECTRON BEAM WELDING OF HORIZONTAL TUBES	Project n° : 15.38/82
Contractor : GERTH/ETPM/ACB/SAF ETPM Address : Courcellor II 33-35 rue d'Alsace 92531 LEVALLOIS PERRET FRANCE Technical director (or person to contact for further information) : Mr ANDRIER	Telephone n° : (1) 759.60.00 Telex : ETPM 612021F

AIM OF THE PROJECT

The purpose of the project was to adapt the method of welding by electron beams to conventional laying of subsea pipelines in view to weld pipes with more elaborate and thicker steel grades.

PROJECT DESCRIPTION

The advantages of the method are its fast welding rate independent of the thickness and minimum change in the methodological characteristics in the molten zone, which remains very narrow and which enables high elastic limit steel to be welded under good conditions.

The difficulties of application of the method lie in keeping the welding zone in a vacuum and in precisely positioning the beam in the plane of the joint of the two tubes to be welded. In addition, since the plane of the joint is vertical, the weld takes place in succession in all positions (flat, up-hand, under-weld and down-hand), thus requiring the adjustment parameters to be varied constantly.

A preliminary study showed that the weldable thickness for an under-weld and a flat weld is limited : the melt falls under the effect of gravity when the thickness is more than 16 mm. Beyond this thickness, the weld must be made in two passes with a non-through beam, one from the inside and one from the outside. Accordingly, two machines, one external and one internal, have to be designed, and the welding method developed, allowing for the results of welding tests made on a simulation bench.

A technical and economic study of the welding method has enabled all the elements involved in its application to be accounted for : handling, setting up on barge, projected productivity, future market study and maintenance.

STATE OF PROJECT

The project began in January 1982 and ended in April 1984.

RESULTS AND APPLICATIONS

The technical feasibility of all the components of the method has been established. However, certain factors must be specified more exactly : weldability of the steels now used for large subsea pipelines, adaptation of existing barges.

Title : VERTICAL POLYPHASIC FLOWS PHASE 1	Project n° : 15.39/82
Contractor : GERTH Address : 4, av. de Bois Préau 92500 RUEIL-MALMAISON FRANCE Technical director (or person to contact for further information) : Mr Gilbert BLU or Mr Bernard MERCIER	Telephone n° : (1) 749.02.14 ext. 2288 or 2747 Telex : 203 050 F

AIM OF THE PROJECT

The objective sought is to develop new methods of computation of pressure losses and flow conditions adapted to polyphasic flows of hydrocarbons in wells and production risers.

PROJECT DESCRIPTION

The study consists of two parts :

Study and testing of diphasic flows of gas and oil in vertical and inclined configuration

This major part resides on experiments under industrial petroleum conditions on 3 inch and 6 inch diphasic test loops (25 metres long), built specifically for this purpose at Boussens. The theories are developed in parallel so as to derive computing models of diphasic flows in a variety of flow conditions (bubble, slug, laminated, annular, dispersed...) and their transition range. Lastly, these models are built into the digital computing programmes of flows in wells and risers in order to account for the numerous physical or thermodynamical parameters involved.

Preliminary study of triphasic flows of gas, oil and water in vertical configuration. A small test loop was built at the Toulouse fluid mechanics laboratory in order to visualize triphasic flow conditions, which are complex and very poorly understood, and to acquire the necessary experimental data for modelization of triphasic flows, which frequently occur in wells, particularly towards the end of exploitation.

STATE OF PROJECT

These two study phases, started at the end of 1981, are progressing normally. They will be completed at the end of 1985 with the development of calculation programmes of the flow in wells producing under natural drive, in risers and lastly in wells employing the commonest activation methods : gas-lift, hydaulic pumping. These calculation programmes will be subjected to tests on actual well data.

RESULTS AND APPLICATIONS

The initial results of these programmes are highly encouraging when compared to the results of conventional calculation methods and measurements on wells.

Title : THERMAL PROTECTION OF TUBINGS	Project n° : 15.40/82
Contractor : GERTH Address : 4, av. de Bois Préau 92500 RUEIL-MALMAISON FRANCE Technical director (or person to contact for further information) : Mr Gilbert BLU or Mr Bernard MERCIER	Telephone n° : (1) 749.02.14 ext. 2288 or 2747 Telex : 203 050 F

AIM OF PROJECT

This project aims at improving the enhanced recovery of heavy oils by steam injection. Its objective is to develop tubing - casing insulation materials enabling the heat losses to be reduced during the steam injection operations while minimizing the risk of breakage of the casing under the combined effect of mechanical and thermal stresses.

PROJECT DESCRIPTION

The project, which is scheduled to cover 1982 to 1984, comprises the choice of the insulation materials and their application techniques, together with experiments simulating actual conditions of use of the materials in order to estimate the performance and cost of the various methods. These materials selected will then be tested on a well in order to evaluate their efficiency.

STATE OF THE PROJECT

The phase devoted to selection of the insulation materials and optimization of their formulae has been completed. The insulation materials thus selected have been manufactured and tested on models of tubing and casing, one consisting of a simplified model operating at atmospheric pressure and the other operating at up to 10 Mpa, so as to stimulate the phenomena linked to the hydrostatic pressure. The first test on a well has been postponed slightly, owing to the present unavailability of a steam injection pilot project and the need to complete the testing of the materials under pressure and during long duration tests.

RESULTS AND APPLICATIONS

Two types of insulating materials have been studied and patented - mineral foams and gelled petroleum oils. Concerning the mineral foams, study of the various parameters had led to the formulation of a silicates mix giving thermally stable foams that have better insulating properties and are more soluble than conventional polysilicate foams.
Study of the various formulae for gelled oils has resulted in the development of a thixotropic composition that can be pumped and be stable up to 300°C, whilst reducing heat losses by convection.
The testing of techniques at a pressure of up to 10 Mpa is currently being continued and the preparation of a test on a well has been undertaken.

Title: Development of a high performance rope system for deep water mooring	Project No.: 15.42/82
Contractor: Société Européenne de Propulsion (Division Propulsion à Poudre et Composites) Address: Boîte Postale No. 37 F - 33165 Saint Medard en Jalles Technical director (or person to contact for further information): Yves Appell	Telephone No.: (56) 34 84 90 Telex: 560678

Aim of the Project:

The mooring with synthetic fibre rope system is an attractive solution, due to the high specific strength, low density and non-corrodable properties of these materials. The purpose of the project is to investigate the feasibility of using synthetic fibre ropes for these mooring systems and to identify the problem areas in order to propose solutions, particularly with regard to the attachment of these ropes, which is the critical item of the system.

Project Description:

The major phases of the programme are:

- documentation research of the existing equipments: materials, rope structures, type of terminations, static and fatigue tests
- state of the requirement: service working conditions, specifications
- synthesis of these design phases: choice of a type of material, rope and termination
- complementary characteristics of the selected rope and termination at a reduced but representative scale (60 tonnes)
- final design of an optimized link system including synthetic rope, termination and structure.

State of Project:

The characterization phase is being performed: fatigue test of

- a rope subject to 60 tonnes
- a rope subject to static tension load and alternate bending.

60 tonne kevlar and polyester ropes are being tested.

Results and applications:

Tests show poor behaviour of kevlar rope with regard to the bending conditions of the test. Link system design is being undertaken in order to minimize bending angle between termination and rope.